Storytelling in Science and Literature

Aperçus: *Histories Texts Cultures*

a Bucknell series

Series Editor: Greg Clingham

Aperçu (apersü). 1882. [Fr.] *A summary exposition, a conspectus.*

Relations among historiography, culture, and textual representation are presently complex and rich in possibilities. Aperçus is a series of books exploring the connections between these crucial terms. Revisionist in intention, Aperçus seeks to open up new possibilities for humanistic knowledge and study, and thus to deepen and extend our understanding of what history, culture, and texts have been and are, as these terms are made to bear on each other by new thinking and writing.

Titles in the Series

Critical Pasts: Writing Criticism, Writing History
ed. Philip Smallwood

History and Nation
ed. Julia Rudolph

Europe Observed: Multiple Gazes in Early Modern Encounters
ed. Kumkum Chatterjee and Clement Hawes

Beyond Douglass: New Perspectives on Early African American Literature
ed. Michael J. Drexler and Ed White

The Patient
ed. Kimberly Myers and Harold Schweizer

Masculinity, Senses, Spirit
ed. Katherine M. Faull

Freedom of Speech: The History of an Idea
ed. Elizabeth Powers

Scotland as Science Fiction
ed. Caroline McCracken-Flesher

Encountering China: Early Modern European Responses
ed. Rachana Sachdev and Qingju Li

Storytelling in Science and Literature

Edited by
Margery Arent Safir

Lewisburg: Bucknell University Press

Published by Bucknell University Press
Co-published with The Rowman & Littlefield Publishing Group, Inc.
4501 Forbes Boulevard, Suite 200, Lanham, Maryland 20706
www.rlpgbooks.com

Unit A, Whitacre Mews, 26-34 Stannary Street, London SE11 4AB, United Kingdom

British Library Cataloguing in Publication Information Available

Library of Congress Cataloging-in-Publication Data

Arent Safir, Margery, 1947-
Storytelling in science and literature / Margery Arent Safir.
pages cm
Includes bibliographical references and index.
ISBN 978-1-61148-643-8 (cloth : alk. paper) — ISBN 978-1-61148-645-2 (pbk. : alk. paper) — ISBN 978-1-61148-644-5 (electronic) 1. Storytelling in literature. 2. Storytelling in mass media. 3. Science—Language. 4. Language and ethics. 5. Narration (Rhetoric) I. Title.

PN56.S7357A74 2015
809'.93353—dc23

2014042067

∞™ The paper used in this publication meets the minimum requirements of American National Standard for Information Sciences—Permanence of Paper for Printed Library Materials, ANSI/NISO Z39.48-1992.

Printed in the United States of America

Contents

List of Illustrations

Margery Arent Safir

Preface

THAT ARGENTINE MAKER OF FICTION WHO GAVE US PIERRE Menard—a Frenchman from Nîmes as author of the *Quixote* and a Kafka who creates his own precursors—had a favorite Greek philosopher. Jorge Luis Borges revisits the fifth century BCE Zeno of Elea in an essay entitled "Avatars of the Tortoise." There, with his usual dose of humor and irony, Borges recalls Zeno's famous paradox about the race between the fleet-footed Achilles and a sloth-footed tortoise who is given a head start, and wherein, by way of a dizzying series of mathematical manipulations, Zeno demonstrates that, try as he might, the swift Achilles will never catch up, much less overtake, the slow-moving tortoise. More spectacular for Borges still is Zeno's argument against movement at all: "Movement is impossible (argues Zeno) for the moving object must cover half of the distance in order to reach its destination, and before reaching the half, half of the half, and before half of the half, half of the half of the half, and before . . ."[1] Later in the essay Borges comes to one of his favorite American philosophers, the pragmatist William James. Two thousand four hundred years after Zeno, James denies that even fourteen minutes can pass: ". . . first it is necessary for seven to pass, and before seven, three and a half, and before the three and a half . . . and so on until the end, the invisible end, through tenuous labyrinths of time."[2]

The mechanism is identical: an infinite division and regression against movement and number that precludes motion in space or in time. The logic is clear, the mathematics irrefutable. Q.E.D. *Except* that the demonstration leads to a conclusion that is utterly false. In the world not of language or number but of matter, Achilles *will* win the race. Movement will occur. Fourteen minutes will pass. Every time. Language that tells the truth and leads to a lie. We are at the frontiers where language and physical matter meet, and frontiers are notoriously perilous zones. As Borges puts it, "it is venturesome to think that a coordination of words . . . can resemble the universe very much."[3]

Zeno's is a paradox for all disciplines. *Qua* paradox, it puts us in the domain of philosophy; its specific problematic (infinitely divisible space and time) places us in the domain of mathematics and—or against—the physical sciences. And since in truth I have not quoted the Greek Zeno at all but rather Jorge Luis Borges's Zeno, we are in the realm of fiction that reaches over time and space, or the discipline of comparative literature.

This paradox, then, introduces the contributors to this volume. Each is a singularly distinguished scholar in one of the disciplines I have just named, yet each is more. Beyond singular distinction in a single discipline each has been venturesome, darting across frontiers, opening crossroads of systems of knowledge and reference, speaking a multiplicity of languages, testifying to a rare ability to converse with other fields. They are people who have taken seriously Buffon's *Discours sur le style*, the appeal of that enlightened naturalist and poet before the Académie française: quite simply, that man make use of *all* his faculties.

Looking at the contributors' biographies and reading their essays, you will see what *used* to be but can no longer so smugly be called the "two cultures," sciences and humanities. Evelyn Fox Keller is a trained theoretical physicist with a thesis in molecular biology, who has taught rhetoric and pioneered work in feminist theory. Jean-Michel Rabaté is a professor of comparative literature as well known for his work in philosophy and in psychoanalysis as for his literary criticism. Theoretical chemist Roald Hoffmann could have been detective Auguste Dupin's ideal criminal—defined as a mathematician and a poet—in an Edgar Allan Poe story famously debated by Jacques Derrida and psychoanalyst Jacques Lacan. Mieke Bal travels continuously across disciplinary boundaries—critical semiotics, the visual arts, transcultural theory, anthropology, narratology, the Old Testament, Rembrandt, popular culture—stopping only to study, as a cultural analyst, the consequences of such "traveling." It is an exceptional gathering of thinkers at the same time rigorous and creative. I am grateful to each for having accepted my invitation to come together around questions that are owned by no discipline alone: language, lies, and ethics, or storytelling in science and literature.

Pure in chemistry, tells Roald Hoffmann, means "unmixed," a neutral description of an element; in theology, *pure* is a moral ideal, a danger when taken to monopolistic extremes; in poetry, *pure* is a defined movement opposed by *impure* poetry; in affective life, *pure* is nonexistent. What "small lies"—or large—creep into the translation from mathematical notation to the murkier world of the word, or from one language to another? When is language conceived of as *by definition* a lie? How to distinguish truth from falsehood, "reality" from "unreality?" To be true, does something have to contain the possibility of being false, to be false the possibility of being true? And how is each defined? By Aristotelian logic? By Galilean empiricism? By Popperian falsification? In "lie," as opposed to "falsehood," a question of intention slips in, thus of ethical behavior, and so the topic not only allows for multidisciplinary reflection, but also poses questions from which no discipline can run away—not least those that are not by name on display here: economics, media and communications, and perhaps foremost in our minds today, politics and international relations, the interactions among nations and peoples, and most particularly when differences of language, value systems, and ethical cultures are present and truth is elusive.

I began with Borges and Zeno and a paradox around language and physical reality. I will end with another, which puts desire in contradiction with matter. In a novel, *La Terre gaste,* French writer Michel Rio evokes "the great totalitarian law"—the law of nature, which is death—to which all living things are subject. The "antibody" to the deadly weight of this law, the freedom from being a slave to it, is the dream of eternity, of never-ending life, which is a fiction. And he writes:

> Language is the organized expression of both the law and its antibody . . . Law-slavery-truth and fiction-freedom-lie, body and antibody, weight

> and counterweight, are as inherent to language as matter and antimatter, or positive charge and negative charge, are to the physical universe.[4]

But the physical universe, as even Borges in his own way recognized, will not be denied. And so Rio concludes: "if . . . life is a fiction that is eternity, and the law of matter the only reality, the cost of the dreamed-of survival of the individual is a suicidal agenda for the species."[5] An ethics not of the soul but of the environment. Language that contains the law and the fiction it provokes, a lie of eternity that soothes the individual at the peril of the species. A lie that seduces, holding out the dream of freedom from the law, the *jouissance* of the freedom of dream; dream, an ethics of desire, an ethics of the unconscious. Thus, we begin: *Storytelling in Science and Literature*.

Notes

1. Jorge Luis Borges, "Avatars of the Tortoise," in *Labyrinths: Selected Stories and Other Writings*, trans. and ed. Donald A. Yates and James E. Irby (New York: New Directions, 1964), 203.

2. Ibid., 207.
3. Ibid.
4. Michel Rio, *La Terre gaste* (Paris: Editions du Seuil, 2003), 59.
5. Ibid., 59–60.

Acknowledgments

THE CONTRIBUTORS TO THIS VOLUME WERE ORIGINALLY brought together for the first annual President's Conference for the Advancement of Scholarship of The American University of Paris, created in May 2003. An earlier version of Margery Arent Safir's introduction appeared in *Connecting Creations: Science-Technology-Literature-Arts* (Santiago de Compostela, Spain: Centro Galego de Arte Contemporanea, 2000); Jean-Michel Rabaté's essay originally written for this volume, "White Lies," was subsequently expanded as "Pieux mensonges: Platon, Nietzsche et Hollywood," in *Tout dire ou ne rien dire: Logiques du mensonge* (Paris: Stock, 2005). A version was translated into English by Suzanne Verderber in *The Ethics of the Lie* (New York: The Other Press, 2007), 142–92.

The original conference papers accurately represent the positions of the contributors within the historical context, and the authors have chosen to reproduce it without change for this publication.

Storytelling in Science and Literature

Margery Arent Safir

Introduction

Beware the Jabberwock[1]: A Cautionary Tale for Interdisciplinary Travel

MANY YEARS AGO, IN THE AFTERMATH OF A NEAR-FATAL accident that left him doubting his mental capacity, Jorge Luis Borges wrote the story I have referred to in the preface, "Pierre Menard, Author of the Quixote."[2] In it, Borges recounts the literary life of a certain Pierre Menard, a twentieth-century Frenchman from the city of Nîmes, who determines to write *Don Quixote.* Adding to the difficulties of this already arduous task, Menard rejects as too facile the idea of merely becoming Cervantes, for whom writing the *Quixote* was in some sense a necessity, and perseveres in remaining a twentieth-century Frenchman, Pierre Menard, writing the *Quixote.* After dogged effort, he manages to produce something. Borges, taking on the role of both barber and priest, scrutinizing Menard's visible and invisible literary production, offers the following comparative reading and assessment of the two *Quixotes.* The first is by Cervantes:

> . . . truth, whose mother is history, rival of time, depository of deeds, witness of the past, exemplar and adviser to the present, and the future's counselor.

Written in the seventeenth century, "written by the 'lay genius' Cervantes," Borges explains, "this enumeration is a mere rhetorical praise of history." Pierre Menard, on the other hand, writes:

> . . . truth, whose mother is history, rival of time, depository of deeds, witness of the past, exemplar and adviser to the present, and the future's counselor.

The Frenchman's words lead Borges to exclaim:

> History, the *mother* of truth: the idea is astounding. Menard, a contemporary of William James, does not define history as an inquiry into reality but as its origin. Historical truth, for him, is not what has happened; it is what we judge to have happened. The final phrases—*exemplar and adviser to the present, and the future's counselor*—are brazenly pragmatic.

Pointing further to the clear stylistic differences between the two authors ("a certain affectation in the archaic style of Menard—quite foreign, after all," whereas Cervantes "handles with ease the current Spanish of his time"), Borges puts forth a "new technique" with which he credits Menard: the art of reading.

What's in a Name?

A word is a word is not a word: Roland Barthes, in a prologue to his *Critical Essays,*[3] demonstrates the short circuits to which communication is prey, taking as his example the word *condolences.* The word means "heartfelt sympathy"; yet if one were to say "condolences" to a close friend who has just lost a loved one, the friend would not hear "heartfelt sympathy" but rather an off-handed, conventional "sorry about that." The word, employed correctly by the "sender," does not communicate its real and correct meaning to the recipient. To communicate the message, the correct word must be adorned and varied; we must "proceed to the frontiers of language," says Barthes. But frontiers are notoriously dangerous. "Be alert, be aware, take heed, take care," we are cautioned in moving across linguistic frontiers from one language to another; be on the lookout for *false friends,* those words that sound friendly to a stranger, that sound like a word one knows, but do not—usually at considerable cost to one's dignity—have the meaning or usage or impact one supposes.

Disciplinary frontiers lurk also. A physicist reading $E=mc^2$ on a page comprehends the equation in the context of a discipline, perhaps of its history; he knows the mathematics involved, understands the homogeneity of equations and why the 2 cannot be replaced by 3, knows the kind of experimentation associated with the theory and the means required to travel from the theory to its justification, in some cases knows the competing theories as well as the implications for later ones. His reading of $E=mc^2$ can never be "virginal." It cannot be identical to that of a layman, more accustomed to seeing the letters and signs of the equation on a T-shirt beneath an image of a tongue-wagging Albert Einstein than in a notebook. Similarly, a musician can explain the term *MIDI* as "Musical Instrument Digital Interface," and some computer-literate listener might even find meaning in the explanation; but the meaning he finds will never be the one understood by someone with a trained ear, who has played an acoustic violin and experienced the evolution of the instrument. Do all physicists think of *Finnegans Wake* when they hear the word *quark*?[4] Do all Joyce specialists think of subatomic particles? Not only does what knowledge we bring to a word matter, but *how* we read is a product, yes, of general culture, but equally of each particular discipline. A physicist and a musician both can use the term *MIDI* and speak of relativity theory—but, as Borges asks in another story, "You who read me, are you sure of understanding my language?"[5]

The Land of "Speaking Corpses"

We are in the realm of *myth*, the term Barthes applies to meaning taken from its context, emptied of its history, or lessened of its original meaning, and reemployed elsewhere.[6] There, in a newly created frontier zone, *meaning* merges with *form*, the two coming together in what he calls "mythical concept." As illustration, we are given a student at a French *lycée* who opens his Latin grammar book to find the words *quia ego nominor leo*, "because my name is lion." In *truth*, the meaning of the words is not intended to tell the student very much at all about the lion, but rather to impose itself on him as an instance of a certain agreement of the predicate. Thus, the ambiguity: as meaning, the naming of the lion already postulates a reading, is a credible whole, complete with a history and an entire system of values; but as the signifier of myth, to make room for the grammatical example, all of that must withdraw to the background. Myth is a type of speech "defined by its intention (*I am a grammatical example*) much more than by its literal sense (*my name is lion*)"; it is "speech *stolen and restored*." The nature of the mythical concept is *appropriation*. Myth is always "language-robbery," playing on the analogy between meaning and form.

But what happens when meaning is too complex or too precise for myth to invade from the outside and take hold of bits and pieces? This is the case of mathematical language, of scientific thought expressed via equation or chemical notation: language that has taken all "possible precautions against *interpretation*." Such language cannot be distorted, but nor is its fate to remain "pure." Where it cannot invade, myth takes away in block, Barthes asserts. Myth does not like holding back, and "the more the 'language-object' resists . . . the greater its final prostitution." Thus, myth takes the selfsame $E=mc^2$ and makes of this "unalterable meaning the pure signifier of mathematicity." Mathematical language itself is a "*finished* language, which derives its very perfection from this acceptance of death." Myth is a language that resists death: "it wrests from the meanings which give it its sustenance an insidious, degraded survival, it provokes in them an artificial reprieve in which it settles comfortably, it turns them into speaking corpses."

Meaning, then, can be looked at, in myth, from two points of view: as the final term of the linguistic system, or as the first term of the mythical system. That ambiguity or that choice of perspective can determine, as it often does in the chapters in this volume, how one tells stories in science and in literature.

Popularization

Specialized material is made accessible to nonspecialists *only* on condition of altering the language used. Popularization means talking about exact technical material (of the "guilds") in the languages common to other fields, for all practical purposes the language of the social sciences and the humanities. A chemist speaking whatever verbal language can communicate with any other chemist anywhere on the planet by drawing chemical diagrams; a mathematician by numbers and formulas; a musician by musical notation. But for the musician to communicate with the chemist, the *lingua franca* of verbal language must be used. Any cross-disciplinary activity involves moving language from one context and employing it in another. It is fundamentally an act of popularization, and popularization is fundamentally a language-based activity; it is, in Barthes's sense, *myth*.

Things get lost, at times broken, in transport, no matter how talented or cautious the mover. A pharmaceutical product taken out of the controlled experimental environment of the laboratory in which it was developed becomes subject to variables ignored in its cloistered world of origin (air pollution, combinations with other products, erratic usage obeying few rules of frequency or dose). A work of art made public is exposed to the likes of Menard or Borges. The uncontrolled encounter is often profitable, yet moving means a loss of control. Transport is a form of translation, and while it can prompt invention in the best sense, its perils are as well known as its treasures. *Traduttore, traditore*: Translation opens up a Pandora's box of questions regarding mishandled meaning, or cracks in the exactitude of meaning; in the gaps interpretation seeps in, and so does appropriation of meaning when the move is from that "finished" language of mathematical equation to the murkier world of the word. If highwaymen roam and hijackers of meaning await there, it is because riches are to be had in movements at the frontier.

But not only is the "natural" language of specialized knowledge altered when taken to other theaters; so, too, is its public. The move implies not only transporting and "translating" the pearls of wisdom but putting them before an audience no longer schooled in how "correctly" to use or appraise them, when to applaud. So delicate an art requires specialists of its own. Thus, enter the storytellers, the purveyors of "upscale popularization"[7] or science as bestseller. These are science writers. John D. Barrow, Jean-Pierre Changeux, Paul Davies, Richard Dawkins, Daniel Dennett, Freeman Dyson, Stephen Jay Gould, Stephen Hawking, François Jacob, Robert Jastrow, Stuart A. Kauffman, Jacques Monod, Roger Penrose, Steven Pinker, Martin Rees, Hubert Reeves, Oliver Sacks, Carl Sagan, Lewis Thomas, Xuan Thuan Trinh, Steven Weinberg, Edmund O. Wilson: a scattering of familiar names suffices as introduction. All are practicing scientists writing in non-scientific language and for educated but not necessarily specialized publics.[8] They are published by commercial presses, as opposed to university presses or journals (where they publish their scholarly work). Their topics range from the earth and life sciences to physics and cosmology to history of science to socio-political essays, and the decidedly nonscientific, even dreamy, titles of a number of their books are illuminating: *Late Night Thoughts on Listening to Mahler's Ninth Symphony* or *The Enchanted Loom* or *The Secret Melody*.[9]

These "storytellers" are the scientists that most of us read. When we read them, we are reading science that is partial in more than one way: we are reading what *can* be popularized, that portion of the whole that can be successfully transported and "translated." Perhaps more importantly, we are reading those people who can communicate in the *lingua franca*, and who wish to do so. This is not every scientist or specialist in a discipline, a fact of which we should not lose sight. And the *us* in "most of us read," the *we* in "we should not lose sight," is a broadly inclusive one, for not only do nonscientists acquire knowledge through scientific popularization, but scientists do as well. The plural is as critical in "sciences" as in "arts." Today, some of the most interesting scientific questions stretch beyond the boundaries of traditional scholarly journals. Yet a paleontologist has no divinely given knowledge of relativity theory or thermodynamics. A result of intra-disciplinary specialization, the "typical" scientific paper is already a popularization of sorts, even when the audience is made up of scientists alone. A rhetorical study of science papers would reveal, for instance, that most begin by giving a history of the problem—its story—and do so because rarely does a scientist address an audience thoroughly knowledgeable about the specific area of research he is explicating.[10]

In this regard, a popular source serves up some matter for thought, or bemusement. Notwithstanding its title—"Unscientific readers of science"—a survey reported in *The Economist* confirms the importance of scientific popularization as a vehicle of information for *scientists*.[11] A first finding: the primary target audience for popularized science is scientists. A second finding: very little is retained by those who read scientific popularizations. The example given is James Gleick's book *Chaos*,[12] a successful popularization by any measure (complete with an inside page of review blurbs entitled "Kudos for *Chaos*"). Clearly presented, it is a "story" well told, and, it seems, actually read by a high proportion of its enthusiastic purchasers. Yet a majority of readers retained from the book only that a butterfly flapping its wings in Miami can cause a storm months later in New York, prompting the unkind suspicion that what one ultimately retains is what mediazation has already fixed in the popular imagination. And this "retention," or lack thereof, held true for scientist-readers and nonscientist-readers alike. Finally, the study offers up one particularly intriguing question. It is clear why scientists might purchase these books, but why do nonscientists, in ever-increasing numbers, purchase scientific popularizations? The answer provided, admittedly not in these terms, can nonetheless be reduced to declaring that nonscientific readers are looking not for science but for "God." Parallel to the increase in sales of popularizing books on science runs an increase in sales of books on mysticism and the occult. Otherwise put, it is not science *qua* science that nonscientific readers are seeking in popularization, but rather almost anything that promises to shed light on the "big questions," the story of life.

Whether informative or speculative, the upscale popularization of science has today assumed certain functions once belonging to other disciplines, not the least among them a central function previously held by literature: holding the "true" keys to the world and its mysteries.[13] Contemporary science has become a veritable mythology of knowledge, and this mythified science is today looked to for clear answers that were once sought in ancient elucidating fictions. The authors of these scientific works have never made claims to such clarity; on the contrary, they as often multiply as resolve fundamental philosophical questions, working with a kind of knowledge that, built upon its own ruins, flaunts the instability of theory and accepts the paradoxes or contradictions of experimental evidence. But it matters little that the legend is contradicted by its own actors. Scientific publication has become the preeminent space for discovery, possesses the strongest hold on denotation, and constitutes the core of modern philosophy by reason of the epistemological and metaphysical questions that, wittingly or not, it raises. If I speak of its popularized form here, it is because travel across disciplinary borders demands popularization, and like all popularization, scientific popularization is a breeding ground for false friends.

Once more, it is Borges who offers up an exemplary demonstration. In an oft-cited essay penned more than a decade after the story of the gentleman from Nîmes, Borges recounts that he has discovered hints of Kafka in the work of Zeno; likewise, in the works of Han Yu, Kierkegaard, and Browning. Reflecting on this odd discovery, Borges concludes that an author creates his own precursors: had Kafka never written, or had Borges never read a page of Kafka, doubtless these disparate authors would never have come together in Borges's mind.[14]

Today, science, too, is creating its own precursors. Through the vehicle of popularization, science is influencing not only the creation but the reception of works of art, in some cases inspiring pages of analysis that sound like Borges stories themselves. The poet Lautréamont has been studied with regard to entropy; Picasso is called "one of the most significant messengers of twentieth-century science";[15] Borges himself is looked at with regard to chaos dynamics. Artists are credited with scientific theories that scientists have arrived at only belatedly, if at all.[16] Superimpositions of this kind make a clear contribution to the art of reading, but are not to be confused with a contribution to knowledge. A coincidence between a certain vision of the world and certain scientific theories alone does not suffice for linkage. Not every work of art that moves from order to disorder is doing so necessarily in recognition of the second law of thermodynamics. This is my caution: beware, if not the Jabberwock, at least the Ettelson effect—that Hasidic doctor, cited by French critic Jean-Jacques Lecercle, who uses Humpty Dumpty's very own methods in the questionable enterprise of proving that *Through the Looking-Glass* is nothing other than a secret Talmudic text.[17]

Taxonomy

In what could be considered a dazzling lesson in taxonomy, Julio Cortázar, in the "Expendable Chapters" of his novel *Hopscotch*,[18] gives us the Uruguayan philosopher Ceferino Piriz, a kind of Linnaeus of the Southern Hemisphere. Organizing the world into phylum, class, order, genus, and species of corporations and ministries, Ceferino offers up such groupings as the National Corporation of Scientific Detectives-Errant and their Houses of Science, The National Corporation of Hospitals and

Related Houses, including not only the expected dental clinics but also workshops for automobile repairs, houses for the pruning of plants, and other such putting-back-in-order activities; or the National Corporation of Commission Agents for Species Colored in Yellow, containing "Generic species colored in yellow or simply yellow: animals with yellow fur, plants with yellow flowers, and minerals with a yellow look," the administrative site for everything pertaining to bananas, "yellow peoples," yellow Volkswagens, and blond hair.[19]

What Cortázar takes ad absurdum, nonetheless, reflects a reality: classification may be arbitrary. Taxonomy can be a question of definition, of choice, of perspective, and of stress. By what criteria related characteristics should be determined was hotly debated in the eighteenth century in which Linnaeus and Buffon worked, and no one today would imagine that classification is isolated from cultural and political context. Taxonomy in its own way puts us face to face with Pierre Menard's "new technique."

To group together is a strategic means of "translating" a discipline by creating new meanings via association; thus, the chapters in this volume necessarily prompt us to look differently at classification. Each contributor writes from the perspective of, yet reaches beyond, the conventional confines of his or her own discipline. How to read these different discourses that nonetheless have common features? Where lie the real frontiers and how are they marked? From what slant to establish the taxons? While not yet a Ceferino Piriz, I have undertaken in this introduction to cut across traditional divisions and to spot shared "traits"—common concerns—in the chapters of these scientists and humanities scholars. What I have talked about as *name*, *myth*, and *popularization* could also be called *modes of imag(in)ing*, of *contextualizing histories*, of *storytelling*.

Who says "storytelling" says "untruth." Whether used to reveal or to mask "real" truth becomes a matter not of science or of literature, but of desire and intent, hence of ethics. *Storytelling in Science and Literature* was conceived in a crossbreeding of the disciplines; that is its story, its history. History may not be the "mother of truth" that Cervantes and Pierre Menard call her, but it is out of a joyous commingling of the sciences and literature that this hybrid volume was born.

Notes

1. The reference is to the poem "Jabberwocky," in *Through the Looking-Glass* (London: Macmillan, 1871) by Lewis Carroll, mathematician and literary author: "Beware the Jabberwock," Carroll warns in brilliant "nonsense verse." In English, *jabberwocky* connotes the transport or fusion of known terms to make new ones, with "nonsense" always a risk of the operation.

2. Jorge Luis Borges, "Pierre Menard, autor del *Quijote*," in *Obras completas 1923–1949* (Buenos Aires: Emecé, 1939; repr. 1989), 444–50. English translation: "Pierre Menard, Author of the *Quixote*," in *Labyrinths: Selected Stories and Other Writings*, trans. and ed. by Donald A. Yates and James E. Irby, 36–44 (New York: New Directions, 1964). Quotations are taken from the English, 43–44.

3. Roland Barthes, *Essais critiques* (Paris: Editions du Seuil, 1964). English translation: *Critical Essays*, trans. Richard Howard (Evanston, IL: Northwestern University Press, 1972). Quotations and summary points are taken from the English, "Preface," xiv–xv.

4. Murray Gell-Mann, credited with the naming of quarks, recounts the origin of the word from his recollection of a passage in Joyce's novel. See *The Quark and the Jaguar: Adventures in the Simple and the Complex* (New York: W. H. Freeman, 1994), 180–81.

5. "An *n* number of possible languages use the same vocabulary; in some of them, the symbol *library* allows the correct definition, . . . but *library* is *bread* or *pyramid* or anything else, and these . . . words which define it have another value. You who read me, are you sure of understanding my language?" Borges, "La Biblioteca de Babel," in *Obras completas 1923–1949*, 465–71. English translation: "The Library of Babel," in *Labyrinths*, 57–58.

6. Barthes elaborates his theory of myth in an early book entitled *Mythologies* (Paris: Editions du Seuil, 1957). English translation: *Mythologies*, trans. Annette Lavers (New York: Hill and Wang, 1972). Quotations and summary points are taken from the English, esp. 115–33.

7. I make this distinction because elsewhere I have discussed the existence of what I call "downscale popularization," or the pure commercialization of science—science not as bestseller but science used for selling best. See my introduction in *Melancholies of Knowledge: Literature in the Age of Science*, ed. Margery Arent Safir (Albany: State University of New York Press, 1999).

8. Gould, Monod, Sagan, and Thomas are deceased. I have limited my list to practicing scientists. Other popularizers, like John Gribbin, who holds a PhD in astrophysics, Simon Singh, who worked as a particle physicist, and Matt Ridley, devote their time fully to science writing, journalism, and television, including the production of science programs. The list of authors cited here, as well as in notes 9 and 10, was compiled at the time of the first publication of this book in 2009. It remains representative of these currents, although many other authors have since also dealt with similar subjects.

9. Lewis Thomas, *Late Night Thoughts on Listening to Mahler's Ninth Symphony* (New York: Bantam Books, 1984); Robert Jastrow, *The Enchanted Loom* (New York: Simon and Schuster Trade, 1981); and Xuan Thuan Trinh, *La Mélodie secrète* (Paris: Fayard, 1988). Within this upscale popularization a subcategory might be reserved for a special class of "speculator scientists." Usually found in quantum physics or cosmology—that is, in disciplines where large quantities of observational data are less available than in others—for some, in thought as well as in title, the frontiers between logic and reverie begin to blur. Some make leaps of faith and, together with philosophers, muse in print as well as in private about the existence of God. See, for example, Horace Freeland Judson, *The Eighth Day of Creation* (London: Penguin, 1994); Paul Davies, *God and the New Physics* (London: Penguin, 1983; rept. 1990); Robert Jastrow, *God and the Astronomers* (New York: W. W. Norton, 1992); and Jean Guitton, Grichka Bogdanov, and Igor Bogdanov, *Dieu et la science* (Paris: Grasset, 1991).

10. A good deal of work has been done on the rhetoric of science. See by way of example Alan G. Gross, *Rhetoric of Science* (Cambridge, MA: Harvard University Press, 1990); *The Literary Structure of Scientific Argument: Historical Studies*, ed. Peter Dear (Philadelphia: University of Pennsylvania Press, 1991); *Understanding Scientific Prose*, ed. Jack Selzer (Madison: University of Wisconsin Press, 1993); Charles Alan Taylor, *Defining Science: A Rhetoric of Demarcation* (Madison: University of Wisconsin Press, 1996); and *Rhetorical Hermeneutics*, ed. Alan G. Gross and William M. Keith (Albany: State University of New York Press, 1997).

11. "Non-scientific readers of science," *The Economist*, May 9, 1998, 97–98. Among those interviewed are Steven Pinker, science writer Nigel Calder, former science editor for Penguin Books Ravi Mirchandani, and Cornell University professor Bruce Lewenstein, editor of *Public Understanding of Science*. Despite the passage of time, this survey remains, in my view, fully relevant today.

12. James Gleick, *Chaos: Making a New Science* (New York: Viking Penguin, 1987).

13. Certain ideas in this paragraph follow Michel Rio, "L'essentiel et l'accessoire," in *Rêve de logique: essais critiques* (Paris: Editions du Seuil, 1992), esp. 83–84.

14. Jorge Luis Borges, "Kafka y sus precursores," in *Otras inquisiciones* (Buenos Aires: Emecé, 1951–1952; repr. 1960), 145–48.

15. Eliane Strosberg, *Art and Science* (Paris: UNESCO, 1999), 162.

16. Michel Serres, among others, points to scientific ideas first "discovered" by literature (*Feux et signaux de brume, Zola* [Paris: Grasset, 1975]), and John D. Barrow looks at non-Euclidean geometrical constructions in *Sriyantra* designs and in fifteenth-century works such as Jan van Eyck's *Arnolfini Wedding* (see *Pi in the Sky: Counting, Thinking, and Being* [Oxford: Clarendon, 1992]; and *The Artful Universe: The Cosmic Source of Human Creativity* [Oxford: Oxford University Press, 1995]). I am not aiming at these or other reliable scholars, but rather at the facile abuse of the tendency to find retrospective precursors and to find them everywhere.

17. Jean-Jacques Lecercle, "Lewis Carroll et le Talmud," *Théorie, Littérature, Enseignement* 10 (1992): 169–87.

18. Julio Cortázar, *Rayuela* (Buenos Aires: Sudamericana, 1963). English translation: *Hopscotch*, trans. Gregory Rabassa (New York: Pantheon, 1966).

19. Ibid., 517.

Evelyn Fox Keller

Can the Circles of Language and Science Be Squared?

MOST SCIENTISTS, EVEN TODAY, CONTINUE TO ASSUME THAT their language, scientific language, is both literal and transparent. Recourse to metaphor is seen as a default option, invoked only when truly scientific language fails them—for example, when attempting to translate their work for public consumption. To call a scientist's attention to her reliance on metaphor in her own work—indeed, in every aspect of that work—is to insult her; for most researchers in the natural sciences, science is to metaphor as truth is to fiction, and rhetoric is another word for lying. But the fact is that scientists cannot proceed without the imprecision that metaphor, polysemy, and other linguistic tropes permit. Making sense of what is not yet known is necessarily a groping in the dark, and for this the imprecision of figurative language is indispensable. Yet, there are also abuses of metaphor in science, and uses of rhetoric that are tantamount to lying. The question I want to discuss is this: How can we distinguish productive from unproductive use of metaphor in science?

Rhetoric and Science

About a dozen years ago, I was teaching at the University of California, Berkeley, with an appointment in the Department of Rhetoric—a choice, I presumed, based on my interest in language and science. So, when a request for proposals for the Ethical, Legal, and Social Issues branch of the then-new Human Genome Project was issued, it seemed entirely fitting for me to respond with an application proposing to analyze "The Rhetoric of the Human Genome Project." Naively, I thought that such a proposal ought to fall well under the Ethical, Legal, and Social Issues umbrella. The scientific reviewers, however, clearly thought otherwise. In fact, they were so enraged by what they took to be my claim that the scientists behind the Human Genome Project were lying, that I was effectively blackballed from those circles for a number of years to come.

Naturally, I was a bit surprised. It had simply never occurred to me that there were academics—especially, I suppose, scientists—for whom the term *rhetoric* would be taken as synonymous with lying. And my immediate thought was that I had better find a way to change departments, and quickly.

But as I now think about this incident in hindsight, I can see the naiveté in my own surprise. In the popular imagination—of both scientists and the lay public—the aim of science is truth; its ethic is honesty, and accordingly, its language is veridical. On the other hand, rhetoric, according to the *Oxford English Dictionary*, is "The art of using language so as to persuade or influence others; [or] the body of rules to be observed by a speaker or writer in order that he may express himself with eloquence."[1] What place, then, could there possibly be for rhetoric in science? Just look at the words: *rhetoric* is in the first instance an art, and it is an art for persuasion and influence; worse yet, for eloquence. Eloquence is a gloss for which we know—just from looking at the professional journals—scientists have little use. If the data speak for themselves, if truth compels, why should scientists have any need for the arts of persuasion and influence? To suggest a need for persuasion would only seem to imply a shortcoming of the data, a deficiency of truth.

Of course, no one really believes in this idealized picture of science. Scientists live in the real world, not in some ethereal, immaterial realm. And they know they must persuade their colleagues of the merit of their work if they are to recruit students, postdocs, get grants, be awarded tenure, or obtain any other of the rewards of academic life. Furthermore, to ensure the availability of funds for research grants, they must persuade and influence congressmen and senators, and, behind the legislature, the public. If it is private funding they seek (more and more common these days, at least in biology), then they must persuade the investors. All this is hardly a secret—everyone knows it, but still, that does not mean it should be made public. Most scientists are realists; they know their own limits, and they know the limits of the popular image of science. They do not exactly believe in this image, but nor do they think of it as a lie. Furthermore, they rely on it. They know that it is persuasive. Perhaps we might even refer to it as *rhetoric*. In fact, scientists are generally aware, at least on some level, of the need to keep many aspects of their venture not exactly secret, but not exactly public either. We might think of these as truths that can cost, and hence, that need to be kept under wraps. Here are some examples:

- Scientists need to sell their work.
- Some are better than others at doing this.
- The often considerable effort scientists invest in ensuring the accuracy and veridicality of their language is far more effective in guaranteeing its close adherence to convention than to veridicality. In fact, I will argue that scientists depend on linguistic imprecision, but of a sort that is clearly bound by convention. In this respect, they are not unlike ordinary language speakers.
- Or, an entirely different kind of example, one that is particularly close to my heart: there is no single definition of a gene on which geneticists can agree; hence, strictly speaking, there is no meaningful way to count the number of genes on a genome. Every molecular geneticist knows this, but it's just not something one wants to say in public.

This lesson, too, I learned the hard way.

Metaphor and the Gene

Let me turn now to the question of metaphor. In rhetoric, a metaphor is a particular kind of figure of speech, but in common parlance it has come to be used for just about any figure of speech. The important point here is that it is not literal; it does not point to the real world with veridicality. Hence, by the popular image of science, it clearly does not belong in proper scientific language. Not quite a lie, but certainly not the talk of truth. More like fiction, perhaps, and certainly less than fact. In my field, the history and philosophy of science, where talk of truth has become distinctly passé, if not a bit disreputable, and a minor industry has developed around the identification of metaphors in scientific discourse, the point of this exercise all too often seems to be a game of "gotcha." That is, even here, the use of metaphor is regarded as somehow disreputable, as discrediting, as confirming a rather lesser view of science than we had been taught. But this is patently absurd. It buys into the very dichotomies we are supposed to have left behind.

A moment ago I promised to defend the claim that science depends on linguistic imprecision, but really, it is just a matter of common sense. Scientific research is typically directed at the elucidation of entities and processes about which no clear understanding exists, and to proceed, researchers must find ways of talking about what they do not know—about that of which they as yet have only glimpses, guesses, speculations. To make sense of their day-to-day efforts, they need to invent words, expressions, forms of speech that can indicate or point to phenomena for which there are no literal descriptors. As Mary Hesse reminds us, "the world does not come naturally parcelled up into sets of identical instances for our inspection and description."[2] Making sense of what is not yet known is thus necessarily an ongoing and provisional activity, a groping in the dark, and for this, the imprecision and flexibility of figurative language are indispensable. And even later, when we think we understand, when we have a theory, Hesse depicts these theories as "models or narratives, initially freely imagined stories about the natural world . . . which depend on a relation of analogy with the real world."[3]

I put forth this argument in my most recent book, *Making Sense of Life*, as a way of understanding both the uses and the productivity of the term *gene* in the narratives of development that geneticists constructed prior to 1953. The gene, I suggested, was:

> not quite a metaphor, or at least not in any strict sense of the term, but a neologism[4] [which acquires] force [by building] on the work of two different metaphors that are between them not only disjoint but in active tension, and conjoins these into a new and an apparently seamless unity.

> Where a metaphor says "this is that," inviting us to see both this and that in new ways, here we have a linguistic construction that amalgamates this and that. It melds into a single form two entities with the disparate properties of atom and organism, and contains the incoherence of such a melding under the protective wrap of a new word. In effect, it offers a resolution of the "riddle of life" by invoking an entity that is a riddle in and of itself.[5]

And I went on to emphasize the productive value of such incoherence. "Without it," I wrote:

> geneticists would have been left without an organizing framework. If the science of genetics was to provide a causal narrative of development, there needed to be a place, a thing, a word to which causal force could be attached. And in the absence of any foreseeable route to clarification of what that thing might be, it was preeminently useful to have a word that could contain or black-box its uncertainty and to keep its internal incoherence, as it were, under wraps. Not only did it permit researchers to get on with their work—resolving the problems they were able to address, without having to worry about those they could not—but also, it provided them with an explanatory framework with which they could "make sense" of the progress they were making in their day to day research, both to themselves and to others.[6]

Linguistic Imprecision and Scientific Productivity

So far so good. But now here's the problem: it has been pointed out to me that I was apparently making just the opposite argument in my previous book, *The Century of the Gene*. There I suggested that the term *gene* has become unproductive because it must now be used to cover so many different usages, that is, that its current imprecision has made it unproductive. The question therefore arises, when is linguistic imprecision productive and when is it unproductive, or even counterproductive? The answer is obviously going to depend, at least in part, on what kind of productivity we have in mind. Sticking with the gene, I want to confine our discussion to the question of scientific productivity—not that there aren't many possible meanings to scientific productivity—there are—and what's more, productivity almost always has its costs. Still, by limiting ourselves to the question about scientific productivity, the discussion is at least somewhat delimited (it excludes, for example, the financial productivity that some see accruing from quite extreme—even outrageous—uses of that word).

I won't pretend to answer this question with anything approaching completeness, but here is at least a part of an answer: that the imprecision and flexibility of the word before 1953 worked so well, so productively, was due, I suggest, precisely to the fact that it worked subliminally. But this possibility ended in 1953 when DNA—a concrete molecule of definite chemical constitution—was identified as the genetic material, and the gene was identified as a specific stretch (or sequence) of that molecule. At that point, the gene became a definite thing. From the usual scientific point of view, this was of course a good thing. But from another point of view—also scientific—the very specificity and concreteness of which the term could now boast deprived it of its conceptual versatility, and hence of its power. In other words, because the gene could no longer be both atom and organism—in fact, it was now explicitly neither—it lost the expansiveness of its former explanatory power. True, it could explain the amino acid sequence of a protein, but it could not account for the development of an organism. A sequence could code for a program, but how could the proteins be arranged to form an organism? From whence would come the directive for producing the required proteins at the right time and in the right place? For this developmental task, a new term—with new sources of tacit imprecision—was required, and the term that was introduced to fill the gap was the *genetic program*. Or, to put it another way, explicit reference to agency in the gene after 1953 would come dangerously close to a demonstrable untruth, whereas, by contrast, the reference of the new term, *genetic program*, was so fluid that virtually no usage could be demonstrated as untrue. Which is not to say that geneticists no longer continued to attribute agency to the gene, but only that they could, at least in principle, be faulted for doing so. On the other hand, no one could be faulted for attributing agency, either explicitly or implicitly, to a notion quite as *toti*-potential as the genetic program.

So let us compare that moment with the situation today. Much has changed since the early days of molecular biology, but the primary change for our purposes here is that the gene lost the univocality of its reference. It no longer refers simply (or solely) to the coding sequence for a single protein—it can also be used to refer to a region of the DNA from which many different coding sequences can be constructed, or even a region from which an RNA molecule is constructed. Why then do I not want to claim productivity for this return of linguistic imprecision? Well, because it is of a very different kind. Multivocality is one thing, and the fluid uncertainty that had earlier accompanied the term altogether another. Rather than having no definite meaning, the problem now is that the gene has an overabundance of definite meanings, each one of which is fixed, as it were, in concrete. Where the conjunction of atom and organism, each *qua* metaphor, required the gene—by virtue of its sheer impossibility—to remain suspended in creative limbo, the multiplicity of specific and unambiguous alternatives that the term now evokes demands resolution. The gene has been made real, and we want to know: Which is it, this molecule or that, this sequence or that? Where is it to be found? Show me.

Nowhere is this problem quite so apparent as in the current furor over numbers. How many genes are there in the human genome? You might say that this is not a very important question, but the point is that once the gene has been made real, it becomes a possible question, and quite a few people seem to want to know how many genes we have. In fact, the question has so much importance to them that it caused quite a stir two years ago when the Human Genome Project announced that human beings have only about 30,000 genes—roughly the same number as the mouse has, and not all that many more than the lowly worm. What's more, many of these genes are so similar that it is possible to exchange mouse genes with human genes and they seem to work quite well in their new hosts. Now, how can this be? The most simplistic reading of these results would seem to leave us with only two choices: either we're not as different from mice and worms as we had thought, or that difference is both real and significant, but it just doesn't have much to do with genes. Either way, we are faced with a pretty large conclusion with equally large implications—all drawn from a claim that turns out, once we recall how many different definitions of *gene* it is possible to employ, to be extraordinarily fragile. Let's suppose, for example, that we ask the counters to use the coding sequence of a protein made by humans, mice, or worms as their operative definition—I don't know the answer, but I would wager that such a count would go a long way in restoring us humans to the elevated status we expect. I think you will have to agree that this is a pretty confusing situation—and it illustrates at least one sense in which the term *gene* seems to have become counterproductive.

New Narratives

We are always told that it is the goal of scientists to pin things down, to make things explicit, to get at the truth. But we are scarcely ever reminded both how fluid things are, and how necessary that fluidity is, on the one hand, for the creative processes that allow us to reach for these—perhaps we might call them "pre-things," that is, for the indefinite constellations that we dimly perceive before we have pinned them down as objects—and on the other hand, for the process of weaving them together into a fabric of meaning. I don't believe that meaning ever emerges spontaneously from assemblages of things, even in physics, but certainly it does not in biology, where the very phenomena that most compel us are those that lend a cell, or an organism, its vitality. Accordingly, the objects we try to pin down in biology always demand a narrative—a story that brings them to life. This seems to be what the folks at the U.S. Department of Energy have now discovered. The very home of the Human Genome Project, having come to recognize that genomes are not themselves alive, has now launched a new project, and it is called "Bringing Genomes to Life." They are not too clear on how they're going to do this, but they have placed their hope in a new pair of words, and in the new kinds of narratives these words suggest, *systems biology*:

> From the foundation of whole-genome sequences, the aspiration of the new biology is to build a new, comprehensive, and profound understanding of complex living systems. . . . The current paradigm in biology—variously described as "single gene," "reductionist," or "linear"—is not likely to be successful on its own. . . . Instead, the existing research approaches will be augmented by a "systems" approach . . .[7]

As I say, new kinds of stories. But lest one be tempted to think of them as "merely" stories, I remind you once again of their productivity, even enormous productivity. The "discourse of gene action" fostered the development of classical genetics and the beginnings of medical genetics; the metaphor of the *genetic program* helped catapult molecular genetics into the most important biological science of the twentieth century; the rhetoric of the Human Genome Project led to the determination of the sequence not only of a human genome, but also of the genomes of a vast variety of other organisms. Last, but not least, narratives of "bringing the genome to life" have already sparked a growth industry in synthetic biology, in creating new kinds of entities inviting the classification of living. In each case, the promises and the myths, whether illusory or realistic, true or false, worked to promote the realization of undeniably real and concrete effects that, in turn, helped foster new promises, new myths, new descriptions, and new rhetoric. And, of course, new effects. That indeed is the inevitable consequence of the human nature of the scientific enterprise.

Notes

1. *Oxford English Dictionary*, 2nd ed., s.v. "rhetoric."
2. Mary B. Hesse, "Tropical Talk: The Myth of the Literal," *Aristotelian Society* 61, supp. (1987): 297–311.
3. Mary B. Hesse, "Models, Metaphors and Truth," in *Knowledge and Language, Metaphor and Knowledge*, ed. Frank R. Ankersmit and Jan Johann Albinn Mooij (Dordrecht: Kluwer Academic Publishers, 1993), vol. 3, 51.
4. Interestingly, *Merriam-Webster's Dictionary* gives two definitions of "neologism": "1. A new word, usage, or expression 2: a meaningless word coined by a psychotic." Available online http://www.merriam-webster.com/dictionary/neologism.
5. Evelyn Fox Keller, *Making Sense of Life* (Cambridge, MA: Harvard University Press, 2002), 131. As Aristotle wrote, "The very nature indeed of a riddle is this, to describe a fact in an impossible combination of words" (Aristotle, *On Poetics*, in *Great Books*, trans. W. Rhys Roberts [Chicago: Encyclopædia Britannica, 1993], vol. 8, 1458a26–27).
6. Keller, *Making Sense*, 132.
7. U.S. Department of Energy, Genomic Science Program, *Notes on the Visions for Computational and Systems Biology Workshop for the Genomes to Life Program*, Washington, DC, September 6–7, 2001, http://doegenomestolife.org/pubs/CompbioVisions-4.pdf.

Jean-Michel Rabaté

White Lies: Plato, Nietzsche, and Hollywood

Can Spies Have a Family Life?

I WILL BEGIN BY CALLING UP THE FIGURE OF THE SPY, WHOSE success in films and mass-market novels is due to a fundamental duplicity, often condensed in the opposition between a "mild-mannered father" by day and an athletic warrior by night, which allows us to imagine at small cost a heroic lining to a lackluster everyday life. The film directed by James Cameron in 1994, *True Lies*, an American version of the 1991 French film *La Totale* (*The Jackpot*), exploits, in a satirical and amusing way, the theme of spies with parallel lives that intersect. Beyond the idea of the double life of the hero caught between his family and his job, it surreptitiously slips in a critique of the first Gulf War, which had recently ended at the time. *True Lies* finds Arnold Schwarzenegger in one of his better roles (the best being that which he is played as Republican governor of the state of California). In *True Lies*, Schwarzenegger proves that he is able to play the classic role of the muscled hero thrown into a series of adventures, all the while getting laughs. In addition, the whole power of the Hollywood fiction machine was required to make the film "work," while the comedy done in the French style—even though it was high-spirited, based on an excellent screenplay, and performed by seasoned actors—was a relative failure. Any meditation on lying is obliged to take into account the delicate "suspension of disbelief" necessary for a "true" lie to be effective. *True Lies*, a film full of sound and fury, moving from abductions to explosions, is a simple remake of the French screenplay by Claude Zidi and Simon Michel. The story does not change in its broad outlines, yet the differences between the two films are striking: Are they due simply to a historical gap (from 1991 to 1994, a new vision of the war between the West and Middle Eastern terrorists emerged), to a change in national perspective (the French irony of a troop of actors who don't take themselves totally seriously, as opposed to the heavy American dream-machine with its entire quasi-militaristic arsenal of special effects), or rather to another manner of dealing with the philosophical problem of the lie?

Schwarzenegger is Harry Tasker, an American secret agent entrusted with perilous missions, and in performing them he deploys the whole panoply of the secret agent, following a tradition that goes back to James Bond: seduction, *savoir faire*, and physical prowess. For seventeen years he has been married to Helen (Jamie Lee Curtis) who, serving as a perfect cover, is unaware of all of her husband's activities. She believes that he is only a sales representative who doesn't talk much and is often away on business. When she becomes seriously bored in her Washington, D.C., suburb between her absent, taciturn husband and her rebellious adolescent daughter, she falls under the spell of a used car salesman who passes himself off as a spy in order to attract her attention. Harry discovers the flirtation and uses the entire arsenal of his special forces team to separate the lovers, still only in the preliminaries. In a discussion, Harry and his rival talk about what women really want. The rival explains that all that's necessary is to make them dream. Harry retorts, "So basically you're lying your ass off the whole the time!"[1] The other opines, going him one better in explaining the degree to which married women are desirable and easy to seduce: it's enough to pretend to be a spy engaged in phony secret missions to get them to fall into your arms. Harry's two worlds meet when he destroys the salesman's love nest to punish him for his indiscretion. Understanding now that his wife dreams of adventures, Harry makes her believe in turn that she can play the role of a secret agent. She agrees to do a striptease to seduce a dangerous spy, none other than her husband in disguise, whom she doesn't recognize and knocks out. At this moment, real terrorists burst in and kidnap both of them. Harry and Helen are brought to a secret base on an island off of Florida where precious Persian and Assyrian statues are found. When cut with a chainsaw, the statues are shown to contain atomic bombs with which the Arab terrorists want to blackmail the American government. If American troops don't leave the Gulf region, a city the size of Miami will be blown up each day. Harry, the secret agent, is the only one who can lend credibility to the blackmail by identifying the bombs. He reveals his true identity to Helen when he gives the names of these "weapons of mass destruction" because he has to be filmed by the terrorists to make their blackmail plot convincing.

The scene of the revelation of his role as a spy takes place in a giant hangar where a beautiful "exotic" spy gives the order to open one of the false Persian statues so as to remove the bomb it contains. (This very well might make us think that if the weapons of mass destruction weren't found in Iraq, it's because they were hidden in the art objects in the Museum of Baghdad looted at the moment of Saddam Hussein's fall.) *True Lies* plays as much on the conventions of Hollywood romantic comedy as on those of action films: what counts for Harry is to gain back the love of his wife, and he finds the key to her heart when he understands that she desires dreams and fiction. He flatters her "Bovaryism" by giving her an active role in his own adventures. When he reveals his identity in front of his wife, he confesses, "What can I say? I'm a spy," in response to which Helen punches him out and yells, "You bastard! You're a lying son of a bitch!"

In the next scene, Tasker is handcuffed to a chair next to Helen, who gradually begins to grasp that her husband has been lying to her all these years. Tasker is injected with a truth serum, and the torturer on duty asks him if there is anything he wants to tell him before he is tortured. Speaking calmly and with great seriousness, Tasker explains that he is going to kill him and describes how he is going to do it, giving the precise details of a series of violent actions. The torturer smirks and asks, "And what makes you think you can do all that?" Calmly, Tasker shows him his broken handcuffs and performs all the actions he described. At the end of an action scene that becomes increasingly pyrotechnic, during which the base is destroyed and hundreds of bad guys are killed, Helen comments, half-admiring, half-exasperated, "I married Rambo." The film closes on a reassuring image: instead of dancing with the beautiful, dangerous spy as he was in the beginning, Tasker is dancing with Helen. Both are enjoying it so much that they pay no attention to the orders being barked into their earphones from the secret agents' van; they have a mission to accomplish! This film contains a few scenes that, viewed retrospectively, seem to be anticipatory moments. In one sequence, an airplane crashes into the windows of an enormous skyscraper; in another, a group of terrorists declares a *jihad* against Americans if they don't leave the Middle East. Can we believe, more than ten years after the film, that the agents accomplished their mission, or will this tango last an eternity?

If the plots are largely the same, one important difference between *True Lies* and *La Totale* is the link to history. In *True Lies*, everything takes place between Washington, D.C., and Florida, but the feeling of the real menace of terrorists from the Gulf launching their "Crimson *Jihad*" is palpable. In *La Totale*, in contrast, history is muted, and the action never

leaves Paris and its suburbs. The couple (played by Thierry Lhermitte and Miou-Miou), kidnapped by terrorists, believing they have been brought to a country in the Near or Far East because they hear Arabic being spoken in the street, discover when they escape that they have only been in Barbès![2] The solution to the couple's problems is the same in both cases: to find again the path of dreams that the humdrum routine of everyday life has effaced. These two comedies examine the foundations of happiness—they are, to refer to the excellent analysis by philosopher Stanley Cavell, "comedies of remarriage."[3] Even if Harry and Helen aren't separated, seventeen years of life in common have carried them to the edge of infidelity. What is therefore at stake is the reaffirmation of the values of the couple. Studying a number of classic Hollywood comedies, such as *The Philadelphia Story*, Cavell demonstrates that what is at stake each time is the American democratic idea that deploys and reorganizes the concept of the "pursuit of happiness" in a given epoch. These comedies are founded on the idea that the first choice was a good one, but it has to be periodically reconfirmed by a second marriage with the same husband or wife.

True Lies goes further and here, for once, Hollywood denounces America's slide into war, even if this denunciation loses its edge in an apotheosis of cataclysmic destruction. Under the effect of the truth serum, Tasker tells his wife everything. She asks, frightened and excited at the same time, "Have you ever killed anyone?" He answers without batting an eye: "Yeah, but they were all bad." The irony is to the second degree because Tasker's naiveté, perfectly incarnated in Schwarzenegger's tense muscles and German diction, is believable, and because it is always possible to mock his pretensions to be on the right side. Nevertheless, let us focus on the different ways the films end. As we saw, *True Lies* ends with the image of happiness shared at last: dreams have become reality for the couple once their daughter has been saved, the telephone call summoning them to new missions is answered by Helen, and the last tango eternalizes this moment of mutual delight that leaves the investigative and political intrigues of the "world outside" in a faraway halo (with the exception of the associates hidden in the van, protesting as usual). True to form, *La Totale* ends on a pirouette: the couple is reunited for an operation in which each one has a role to play, everything falls into place perfectly, the agents succeed in capturing the dangerous terrorist; but in a final reversal, it is none other than Michel Boujenah, the fake spy, who has fooled everyone once again.

When everything becomes comedy, even war and terrorism, the happy ending reaffirms the values of the couple through a parallel reversal that closes the film. In one case it is believable, in the other it is not. Not one second of the sequence in which Thierry Lhermitte mobilizes his entire team of spies to find his son, who cuts his classes to play with a rap group in an abandoned factory in the suburbs, is believable. By contrast, one can get completely caught up in the scene where Schwarzenegger saves his daughter, who has been taken hostage by terrorists in Miami, flying to her aid in a military jet and destroying the entire gang, a part of the city, and a helicopter along the way; humor (Tasker yells "You're fired!" to the head of the terrorists while launching him on one of his jet rockets to strike the helicopter of the other terrorists) does not interfere with the instinctive participation of the spectator. Undoubtedly, Schwarzenegger's physique, his accent, and his whole bodybuilding brio were required to make all this happen at the same time. The adventure remains credible thanks to him; he manages to make us take the stereotypes seriously, particularly when they denounce themselves.

The obligatory lie of the spy who cannot tell the truth under pain of death, and thus who cannot blow his cover to his family, comes up against the gratuitous or voluntary lie of those who, tempted by adultery, find in deception the conditions for a successful "adventure." The explosive encounter between Madame Bovary and James Bond allows us to understand, retrospectively, that one of the driving forces of Emma's tragedy was to never have awakened the suspicions of her idiot husband. When one and the other of the two deceptions are unmasked, it is the myth of the double life that explodes: the wife who was a cover is invited to share in the heroic exploits, and the secret agent no longer has to lie to his family to complete his missions; espionage becomes a family affair. It is clear how romantic comedy (romance) assures social cohesion on a mythical level: it permits the reinsertion of the individual into the couple and at the same time into the collectivity. However, a lie will be said to be "true" if one can believe in the fiction all the while knowing that it is a fiction. It is on this level that Hollywood can deploy the means that force credibility, or at least make it enter into the cliché codes of war movies—a ballet of real airplanes and helicopters in the sky, atomic bombs that explode, and buildings that collapse. By contrast, the troop of French actors, often gathered together in excellent comedies, is not striving to force credibility; it's obvious that they don't believe the story—their winks are everywhere. What counts for *La Totale*—whose title declares a healthy skepticism for the exploits of supermen who are spies, supermen capable of succeeding in military exploits but vanquished in their weak spot, their private lives—is to be enjoyable entertainment. As a contemporary comedy of remarriage, *True Lies* poses more intensely the question of the necessary relations between political deception, sentimental subterfuge, and entertainment, all while reflecting on the Hollywood fantasy machine. If, as we have seen, the lie is not "true" in *La Totale,* while it is certainly "true" in *True Lies,* it is because an entire machinery of lying is required. An artistic lie can be said to be "true" as long as it is not content to just "totalize" the fundamental givens of the relations between private life and social life by exaggerating them humorously (*C'est la totale!*—"It's the jackpot!"—Eddy Mitchell says to Thierry Lhermitte). But it also presents a serious question: Is a successful work of art a "true" lie in the sense that it forces us to preserve our illusions? If all truth initially presents itself as a necessary lie, is it necessary to approve of the existence of these lies, short of limiting their production, or rather to condemn them? The urgency of reopening a fundamental philosophical debate on the founding or primary lie courses through this double film, *La Totale* and *True Lies*, a debate that opposes the arguments of Plato to those of Nietzsche. It thus makes sense to examine them before testing our hypotheses on a more recent Hollywood film, *Troy*.

Nietzsche and the Fiction of Truth

In the successful adaptation of *La Totale* into *True Lies*, have we been witness to a Nietzschean apology for "extra-moral" lying under the cover of entertainment? It is Nietzsche who has systematically attacked our belief in the truth, and the impact of his youthful essay entitled *On Truth and Lying in an Extra-Moral Sense* (1873)[4] was immense in the 1970s in France and the United States when it was exhumed from unclassified posthumous papers and translated.[5] This key text appears in all of the American anthologies of what is called *theory*. For Nietzsche, the point of departure for the analysis of the Platonic and Christian metaphysical appropriation of the truth can only be art—initially conceived as Greek tragedy and Wagnerian opera—which will be defined as a repertoire of necessary lies. This is what he writes in a text dated the summer of 1873:

> Art. Necessary lies [*Notlüge*] and voluntary lies [*Freilüge*]. The latter still must be derived from some form of necessity [*Not*].
>
> All lies are necessary lies. The pleasure of lying is artistic. Otherwise, only truth has pleasure in itself. Artistic pleasure is the greatest kind because it speaks the truth quite generally in the form of lies.
>
> Concept of personality, indeed, that of moral freedom as well, are necessary illusions, so that even our drives for truth rest on the foundation of lies.[6]

The German expression that Nietzsche develops is *Notlüge*, which means a necessary—or "white"—lie, that "official" or "officious" lie, a fabrication demanded by a situation of distress (which is one of the meanings of *Not*) or an "aid" in the case of pressing need (as one speaks of *Notbremse*, an "emergency break"). There are no longer differences between tendentious or "black" lies and *pious* or "white" lies (the English language not so long ago opposed *white lies*, judged to be inoffensive, to *black lies*, judged to be malevolent and perverse; the first did good, the second evil—as is evident, the vocabulary has retained traces of centuries of moral casuistry). With Nietzsche the philologist, morality can no longer pass its casuistic razor between these two notions because morality itself is founded on a lie. Does that then mean that "everything lies"? No: art, at least, does not lie, in the sense that it knows itself to be an illusion. In pointing the finger at itself, it henceforth denounces contemporary illusions. Another extract develops this point:

> How is art as lie even possible!
>
> When closed, my eyes see within themselves countless and changing images—imagination produces them, and I know that they do not correspond to reality. Thus, I believe in them only as images, not as realities.
>
> Surfaces, forms.
>
> Art includes the joy of awakening belief by means of surfaces: but isn't one really being deceived? For if so, it would cease to be art!
>
> Art's aim is deception—but we are not deceived?
>
> What is the source of the pleasure we take in the attempted deception, in the semblance that is always recognized to be semblance?
>
> Art thus treats semblance as semblance, precisely does not want to deceive, is true.[7]

Truth consists in revealing the mask; the mask stripped naked leaves behind it a useless network of moralizing calculations on the necessities of state, pressing obligations by which an attempt is made to make an excuse for lying. Art does not forgive; it never apologizes, even when it denounces itself as a fiction. It develops the tendency toward metaphor that is found at the foundations of language, poetry, dreams, and myths. Nietzsche takes up the intuition of Giambattista Vico, who defined myths as "true stories," but also as metaphors that required interpretation. Thus, according to Vico, the image of Jupiter brandishing lightning is neither a poetic exaggeration nor a symbol of his power, but rather an ancestral memory of the path followed by humanity, beginning in the archaic epoch, when the first men were wrenched from a practically animal state by religious fear of thunder, a natural phenomenon they did not understand but took for a divine sign, the commandment of a stern father ordering them to humanize themselves a little.[8]

The truth Nietzsche is speaking about is less historical than that of Vico, but no less interpretative: recall that these notes are contemporaneous with the first *Untimely Meditations*, which put forth, as a rule, an attack against the historical outlook and antiquarian mentality of the German university. One of the constant themes of Nietzsche's work, which he develops more systematically in his later years, is the idea that the truth is only an interpretation, that there are no objective "facts" that can be dissociated from an interested perspective. The critique of myth will thus be devoted to bringing to life this subjective and often passionate "interest."

In the years 1873–1876, Nietzsche began to take a distance from Schopenhauer, and then from Wagner, and when he attacks teaching institutions in Germany, he does so by putting forward the model from pre-Socratic and classical Greece. How can the fictions of art be accepted as another level of truth? The lesson of *On Truth and Lying in an Extra-Moral Sense* is twofold: on the one hand, the truth cannot be attained through conventional morality, and on the other, everything is language, and language consists of multiple traces of effaced metaphors (a theory that also follows from Vico). In addition, the essay contains a fairly regressive reverie on a mythical ideal state identified with the polytheistic religions of Greece and Rome. Nietzsche becomes lyrical in this passage from the second part of the essay:

> When any tree may begin anytime to speak as a nymph, or a god in the guise of a bull can abduct a maiden, when the goddess Athena herself is suddenly seen driving through the marketplaces of Athens on a beautiful team of horses in the company of Pisistratus—as the honest Athenian believed—then at any moment as in a dream, anything is possible, and all nature crowds around man as if it were only the masquerade of the gods, who only make a joke of deceiving man in all forms.[9]

What is surprising in this passage is that Nietzsche, who aims to destroy a thousand-year-old tradition in order to later "philosophize with a hammer," here positions himself on the opposite pole from a more skeptical Hellenistic critique directed at the sincerity native to the Athenians.

This critique, of which Plato often provides the model, aimed to deny all dignity to the mythological "lies," to the jumble of anthropomorphic legends trafficked by Homer and the poets. If these stories of metamorphoses, abductions, or apparitions were not interpreted allegorically or metaphorically, they were only to be seen as stories for naïve children. One of those who was accused of being an "Antichrist" well before Nietzsche was Lucian of Samosata, the Greek rhetorician of the second century AD who made a practice of believing neither in the myths of the classical Pantheon nor in the ideas spread in the declining Roman Empire, leaving an opening for nascent Christianity. In one of his overwrought rhetorical passages, entitled "The Lover of Lies," he mocks the credulity of the Athenians of the time. This dialogue on the propensity to believe gibberish about miraculous cures, phantoms, prophetic dreams, the resurrection of the dead, enchantments—in short, about all the latest superstitions or multiple stories about the supernatural, which Lucian treats uniformly as "lies"—does not even spare the historians: Herodotus is cited next to Homer as being among those who had the audacity to leave their lies in writing "so that they not only deceived those who listened to them then, but transmitted the falsehood from generation to generation even down to us, conserved in the choicest of diction and rhythm."[10] In the final analysis, Lucian's satirical skepticism rests less on a critique of content (most of the fantastic stories he treats as lies are otherwise so unlikely that they are barely believable for a modern reader) than on a subtle analysis of a psychological tendency, the need to believe in the supernatural and in "other worlds" in the name of which one is ready to silence one's reason or good sense.[11] What Lucian's ironic rationalism permits us to apprehend are the last glimmerings of the pagan world before its conquest by Christianity, an urban skepticism, at the same time sophisticated and disabused, which to a degree evokes our own.

It will take Nietzsche more time to reconcile with Voltaire and the "French" and "modern" mockeries directed at "lies" trafficked by belief in the occult. This is no doubt because he had not yet broken with a certain mysticism of art, inherited from Schopenhauer, who maintained his belief in prophetic dreams and premonitory apparitions, and even endeavored to confer a new philosophical dignity upon them. In fact, during this period, Nietzsche was not at all opposed to the world of dreams: "Actually, the wide awake person is certain that he is awake only because of the rigidly regular web of concepts, and so he sometimes comes to believe that he is dreaming when at times that web of concepts is torn apart by art."[12] He goes as far as citing by memory (though slightly distorting) Pascal's celebrated apology: an artisan who dreams every night that he is king will be just as happy as a king who dreams every night that he is an artisan. At the very least, this proves that it is necessary to take the unconscious into account as soon as lies are at issue, and on this point Nietzsche, more than Schopenhauer (whose system, by contrast, rests on an unconscious Will that traverses Nature), opens the way to Freud. In addition, for Nietzsche, the problematic of the lie opens onto the great theme of the classical moralists, the coefficient of illusion that accompanies any judgment in the sense that one always lies to oneself. It is with this great French tradition, that of Pascal, La Rochefoucauld, and Chamfort, that Nietzsche becomes reconciled in his final years. But in 1873, Nietzsche endeavors to show himself to be still more "modern" than his predecessors like Schopenhauer, who at base was only a Platonist hiding beneath a Kantian mask.

Nietzsche makes critical revisions of his own doctrine, the sign of a constant "modernization" in process. This idea can be seen to reappear in the 1885–1886 notebooks, as well as in the "Attempt at Self-Criticism" of 1886 that accompanies the second publication of the *Birth of Tragedy* (published for the first time in 1872). In this curious palinode, Nietzsche critiques his former pessimism and his romantic excesses that approached, he says, *Sturm und Drang*. He realizes that in his rereading of the tragedies in the name of the Dionysos-Apollo opposition, superimposed exactly on the Schopenhauerian idea of the Will-Representation opposition, he did not pay enough attention to that new figure of "objective" truth provided by science. In 1886, he grasps that the new European thought accords a blind confidence to science by way of a latent or overt positivism that causes all the phantoms inherited from myths, religions, and popular superstitions to fade. "What I had got hold of at that time was something fearsome and dangerous, a problem with horns, not necessarily a bull, but at any rate a new problem; today I would say that it was the problem of science (*Wissenschaft*) itself, science grasped for the first time as something problematic and questionable."[13] Nietzsche treats the problem of truth and lying not like a matador who kills the bull and cuts off its horns, but rather like one of those Cretan acrobats, of which we have iconographic traces, who executes actual ballet moves and pirouettes over the bull's horns.

It is thus that Nietzsche discovers that the perspective must be more than double, that in order to avoid being caught in the horns of a dilemma it is necessary to displace the parallax toward a third term: it will henceforth be necessary "to look at science through the prism of the artist, but also to look at art through the prism of life."[14] In crying "God is dead!" Nietzsche is not only deploring the loss of a transcendental signifier or of a stable and solid foundation for the universe of discourse. He wants above all to say that the power to create fictions (since there is only that) must revert to men themselves and is not to be left to God. The God of which he speaks will be closer to Dionysos or even Apollo than to the Christian God; he is a tragic god who, when he creates or destroys, is only ruled by his own pleasure. Nietzsche will thus be able to sign his final letters, said to be written while he was insane, indifferently with the name of "Dionysos" or of the "Crucified."[15]

From the Necessary to the Organic Lie

If everyday life is desperately banal because it is too "quotidian," as Jules Laforgue humorously lamented, the simple possibility of holding in memory two different versions of everything that happens provides a good exercise for would-be creators, a useful gymnastics of the mind, a spice added to the narrations one can make of them. For everything happens in language—what counts is certainly not the consciousness retained of various events, but the possibility of transforming those events into stories. If the zest of moral shame is added, the trick has been played: creative interiority is combined with a subversive, even Satanic edge. The habit has been set in place, the pose is ready. When I no longer have these two discourses at the same time, I fall into another danger: that of believing in my fictions, and thus of making myself divine, and sharing Nietzsche's tragic lot. Even as it would be absurd to accuse a psychiatric patient who claims he was abducted by aliens of being a liar, in the same way, my "truth" implies the permanence of a fictive lining that fabricates simultaneously at least two discourses about me.

This becomes dangerous at the moment when creative duplicity becomes a principle of political government or when it gives birth to a morality for all. As soon as the act of lying is analyzed, it entails entry into a certain form of casuistry, that is to say into more or less sordid calculations on intentions and wills that do not coincide with acts—which brings with it the deployment of all the pretenses of extenuating circumstances. Nietzsche, who opposes the necessary lie—the white lie—to the free lie of art, does not escape this stumbling block. Pascal thought he had escaped it, hoping that taking recourse in a rigid faith, guided by God, would offer a certain value against all of the shams of the Jesuits whose casuist quibbles flattered the hypocrisy of their flocks. For Max Scheler, by contrast, Pascal appeared to be the man of *ressentiment par excellence*. Who is the man of *ressentiment*? It is he whom Nietzsche accuses of having the mentality of a slave, he who cannot "affirm" or fully enjoy life but needs the protection of social, moral, and religious codes to protect himself from others, and proceeds to attack all those who "affirm" and who therefore seem to him to be too free. Faithful to the lineage of Nietzschean thought, in his book on *ressentiment* Scheler describes how he is more closely allied with lying than with perversion. And what he calls an "organic" lie in fact approaches the idea of an unconscious lie:

> Beyond all conscious lying and falsifying, there is a deeper "organic mendacity" [*organische Verlogenheit*]. Here the falsification is not formed in consciousness, but at the same stage of the mental process as the impressions and value feelings themselves: on the road of experience into consciousness. There is "organic mendacity" whenever a man's mind admits only those impressions which serve his "interest" or his instinctive attitude. Already in the process of mental reproduction and recollection, the contents of his experience are modified in this direction. He who is "mendacious" [*verlogen*] has no need to lie![16]

Such a "lie," transformed into instinct and vital deformation, is not distinctly different from the Freudian unconscious. In closely analyzing the stories of those who allowed themselves to be taken in by a generalized "lie," we have seen that they were also voluntarily deluded, and these delusions were transformed into life programs engendering false promises. The lie to the other almost inevitably doubles the lie to oneself.

From Nietzsche to Plato

This redoubling produces a vicious circle, for if one systematically lies to oneself, is it still possible to speak of "lying" in the sense that the latter presupposes at least two parallel discourses founded on a knowledge of the truth? If one lies to oneself in a passive, "organic" lie, can one still be held responsible for the deceptions imposed by the collectivity? There indeed exists something resembling a generalized duplicity produced by culture, by our civilization that promises a rosy future to a suffering humanity—this is at least Freud's point of view in *Civilization and Its Discontents*. Nietzsche is never far from this pessimistic vision of the fundamental "sickness in being" of men and women. However, he will not necessarily be in agreement on all points with his disciple Scheler, at least in 1873, in the sense that he does not set the organic lie in opposition to a truth that is full, physical, and corporal. Nietzsche knows that the truth, like the lie, is often out of reach, both one and the other stricken with an original repression that necessitates an obligatory and substitutive lie. He thus distinguishes, very usefully, veracity (the will to say the truth) and truth:

> Truthfulness, as the foundation of all compacts and the prerequisite for the survival of the human race, is a eudaemonistic demand: it is opposed by the knowledge that the supreme welfare of human beings lies rather in illusions: that is, that according to the fundamental principle of eudaemonism, truth and lies would have to be employed—as, in fact, is actually the case. Concept of the forbidden truth, that is, a truth whose function is to conceal and disguise precisely the eudaemonistic lie. Opposite: the forbidden lie, which occurs in the realm governed by permitted truth.
>
> Symbol of forbidden truth: *fiat veritas, pereat mundus.*
>
> Symbol of forbidden lie: *fiat mendacium! pereat mundus.*[17]

In this dialectical analysis of binary opposites, Nietzsche does not conclude too rapidly because he distrusts any judgment that would make everything tip either to the side of skepticism or to the side of the Darwinism of triumphant science. The "forbidden truth" affirms: "Let the world perish as long as it is the truth!" The "forbidden lie" retorts: "Let the world perish as long as it is a lie!" This double absolutism comes into conflict with eudaemonism—that is, with theories that affirm that the aim of life is happiness. Nietzsche concludes this fragment in a relatively Schopenhauerian fashion through a rare affirmation of the combined power of love and art: "The truest things in this world—love, religion, and art. The first of these sees through all dissimulation and disguises, it penetrates to the core, the suffering individual, and suffers along with him; the last of these, as practical love, gives consolation for this suffering by telling about another world order and teaching us to disdain this one."[18] Nietzsche knows what he is talking about when he accuses, a bit later, the mystical aesthetes, followers of Schopenhauer, of being "denigrators of the world."

It was necessary to undertake a detailed analysis of the fundamental Nietzschean concept of the necessary or white lie (*Notlüge*), for we will see it return in Nietzsche's reading of Plato, or rather in his interpretation of the place of the lie in Plato. In the final pages of the second "Untimely Meditations" that deals with the "Uses and Disadvantages of History for Life," Nietzsche returns to the fact that Plato founds his entire system of government on a "necessary" lie—a lie that is both forced and "white" at the same time. He cites the famous passage from the *Republic* that exposes the "noble lie," a fiction that Plato posits as a necessary moment, a deception that will make possible the solidification of the social body of his ideal city. For Nietzsche, this political fiction is the paradigm of the forced lie that defines politics in its entirety: "Plato considered it necessary that the first generation of his new society (in the perfect state) should be educated with the aid of a mighty necessary lie [*kräftige Notlüge*]: the children were to be taught to believe that they had all formerly dwelt asleep under the earth, where they had been kneaded into shape by nature's workman."[19]

Let us return to the passage in the *Republic* to better understand what is at play. Previously, two times at least, the question of the lie had been broached in its complexity. In a long discussion of falsity, book 2 condemns the lie as absolutely detestable,[20] and, when he examines the "fictions favorable to virtue," Plato sharply criticizes Homer and the other poets who often depict the gods in an undignified or childish light. The "falsehoods"[21] of mythology will thus be strictly controlled. Among the arguments that sprout up is the fact that the poets do not hesitate to depict the gods as liars:

> But, said I, may we suppose that while the gods themselves are incapable of change they cause us to fancy that they appear in many shapes deceiving and practicing magic upon us?—Perhaps, said he.—Consider, said I. Would a god wish to deceive, or lie, by presenting in either word or action what is only appearance?—I don't know, said he.—Don't you know, said I, that the veritable lie, if the expression is permissible, is a thing that all gods and men abhor?—What do you mean? he said.—This, said I, that falsehood in the most vital part of themselves, and about their most vital concerns, is something that no one willingly accepts, but it is there above all that everyone fears it.[22]

38 The conclusion of this development is that gods as well as men detest the lie.

In book 3, it is the poets who are directly blamed for the excesses and distortions that are harmful to the education of children. Accordingly, it will be out of the question to show the gods roaring with laughter and other similarly inappropriate images. However, in the course of his argument, Plato suggests that the lie must be confined within strict limits, but that it can prove useful to rulers.

> But further we must surely prize the truth most highly. For if we were right in what we were just saying and falsehood is in very deed useless to gods, but to men useful as a remedy or form of medicine, it is obvious that such a thing must be assigned to physicians, and laymen should have nothing to do with it.—Obviously, he replied.—The rulers then of the city may, if anybody, fitly lie on account of enemies or citizens for the benefit of the state; no others may have anything to do with it.[23]

Next, Plato reviews a series of extracts from Homer, which are funny, disquieting, and of often questionable taste, which deserved to be censured so as not to influence the imaginations of malleable young children; Plato legitimates the State lie, all the while preparing the ground for the theory of the "noble" lie that is to follow. More precisely, the *Republic* justifies the founding lie of the community. It concerns a lie directly targeting the rulers and soldiers of the republic, to whom will be addressed (the discourse is projected in the future) the following harangue:

> And yet I hardly know how to find the audacity or the words to speak and undertake to persuade first the rulers themselves and the soldiers and then the rest of the city that in good sooth all our training and educating of them were things that they imagined and that happened to them as it were in a dream, but that in reality at that time they were down within the earth being molded and fostered themselves while their weapons and the rest of their equipment were being fashioned. And when they were quite finished the earth as being their mother delivered them, and now as if their land were their mother and their nurse they ought to take thought for her and defend her against any attack and regard the other citizens as their brothers and children of the selfsame earth.[24]

This new myth of origins frightens the interlocutor: "It is not for nothing, he said, that you were so bashful about coming out with your lie,"[25] and he admits a bit later that there is no chance that the story will convince the first generations.[26] What is at issue is a myth of absolute autochthony that transforms state membership into familial membership. At the same time, it naturalizes the different statuses of citizens, whose souls are stamped with a type of metal: gold for philosophers and rulers, silver for the soldiers and helpers, and iron for the artisans and farmers. No one must rebel against his social status. If all the fellow citizens are "like" brothers and sisters, the generalized risk of incest exists. However, Plato is willing to take this risk as long as it assures the cohesion of the social body: "And as you are all kin, though for the most part you will breed after your kinds . . ."[27] Marc Shell has admirably commented on the diverse national myths that transform the affairs of the political community into a blood community (and, in this story, the land is not far, as if to summon from afar the Nazi myths of *Blut und Boden*) as well as on the theories of universal fraternity that always end by justifying incest.[28] As he demonstrates, any apparently generous theory of universal fraternity—we are all brothers and sisters—leads to a logic of deadly exclusion: since it seems impossible to deny that some among them are not my "brothers" (those who do not resemble me, for example), it is therefore necessary to conclude that they are not "human." It is in the name of this cunning paralogism, of this lying, murderous sophism, that humanist universalism will come to transform itself into a machine of death that will seek to eliminate, by all possible means, those who will be treated as "subhuman," or, worse, as animals.

What is surprising about the Platonic myth of the autochthon—recall that an "autochthon" is literally he who is born from the earth, as opposed to an immigrant or any sort of foreigner—is the poetic way that Plato literalizes the earth. To believe it, he says, it is necessary only that the children think that they were born under the earth. To become the earth-mother, the earth must absolutely replace the mother. No more mothers, no more fathers, only a gigantic natural cauldron which will fashion the citizens to come. In the absence of such a belief, there is always the risk of emphasizing different roots, other origins, or exogamous or exogenous filiations. This leads Plato to dramatize the rhetorical difficulties of his own story, insisting on the narratological paradoxes produced by the "beautiful necessary lie." Besides, he adds, as if to enhance his credibility, the myth has no pretensions to originality: coming from Phoenicia, it issued from a long tradition and spread into several countries. But he mentions no certain source: the origin remains as vague as a rumor that is, in addition, contaminated by the stories of the poets (Plato adds "as the poets aver and have induced men to believe"[29]). Why must he present this political fiction as a lie in which no one really believes, neither the orators who are present nor future listeners? And why must it be ennobled as much by its salvational intentions as by its exotic origins? This is not an isolated example: in numerous dialogues, Plato takes recourse to this same salvational lie, a lie justified by a hypothetical result, which would be to render children more just and later turn them into good citizens.

Thus, in *Laws*, the deceitful story evolves, appropriating the Phoenician myth of Cadmus, who plants the teeth of a dragon in the earth and harvests armed men. According to the Athenian, the legislator absolutely has the right to invent a "lying fiction" of this kind, as long as it is aimed at the well-being of the youth.[30] If it happens that a community comes together and speaks "in one and the same voice," it will have won and have no need to be ashamed of its subterfuge. Similarly, in *Statesman*, the stranger who speaks to Socrates seems convinced that this myth of origins is well-founded and regrets that it is no longer accorded belief:

> Our legends tell us that once upon a time there was an earthborn race. Now it was this race which at that moment of crisis began to return to life out of the earth. The memory of it has lived on, for it was handed down to us by the earliest of our forbears. These earliest forbears were the children of earthborn parents: they lived in the period directly following the end of the era of the earthborn, at the close of the former period of cosmic rotation and the beginning of the present one. These ancestors of ours passed on to us these stories of the earthborn, and it is an unsound judgment to disbelieve them as so many do nowadays.[31]

Myths serve several functions in Plato's system, and it would be forcing things to fit them all into the category of "noble lies." However, as far as the educational myth is concerned, one which implies ethical rectitude in conformity with a pedagogical program as a duty, it must not be forgotten that it is only a fiction in which belief itself is problematic.

The Imposture of the Adversary and the Crafty Odysseus

Is it necessary to force oneself to believe in lies, even beautiful and noble ones, if one knows that they are lies? Could there be an ultimate lie to which, by a sort of heroic leap of imagination, one could finally adhere? By these literally discordant questions I would like to conjure the image of a Plato who is practically Nietzschean, of a philosopher who is less sure of possessing the truth than he appears. In the same way as Plato multiplied the *effets de cadre*—what Jacques Derrida calls the *parergon* in *Truth in Painting*—and the dialogues of diverse interlocutors, and created whole cosmogonies to justify questionable politics, Nietzsche spun out his last myth—his last desperate belief, a faith in the eternal return of the same—as his own "noble lie." It was necessary to attempt to believe in it: the thought of the return of the world as it is now, for always, would, he admitted, drive him mad, had already driven him mad. The thought of the eternal return was to be a joyous and positive affirmation that would transform life at the same time into an event heavy with meaning and into a pure spectacle, since everything would be nothing other than a pure "repetition"[32] in the theatrical sense as well.

Like Plato, Nietzsche knows that all pedagogy is founded upon imposture. Here is how he puts it in *Human, All Too Human*:

> With all great deceivers there is a noteworthy occurrence to which they owe their power. In the actual act of deception, with all its preparations, its enthralling in voice, expression and gesture, in the midst of the scenery designed to give it effect, they are overcome by belief in themselves: it is this which then speaks so miraculously and compellingly to those who surround them. The founders of religions are distinguished from these great deceivers by the fact that they never emerge from this state of self-deception: or very rarely they experience for once that moment of clarity when doubt overcomes them; usually, however, they comfort themselves by ascribing these moments of clarity to the evil antagonist.[33]

Did Nietzsche go mad for suddenly believing in himself, causing himself to be duped by his own white lies? At this moment, he would thus have become himself and his Adversary, tearing himself apart in this mortal embrace.

We have seen what the power of the Adversary can be once it has seized the imagination of the subject. It is better for the moment to place confidence in the truth of illusion. But still, undoubtedly, we do not know enough about the relationship between truth and lying. To go further, I propose to interrogate Plato in one of his earliest dialogues, undoubtedly the first, the *Lesser Hippias*, a short and simple dialogue in terms of its argumentation, which is consecrated to the "False." Socrates had just come from listening to an eminent sophist, Hippias of Elis, speak on the subject of Homer. In a comparison between Achilles and Odysseus, Hippias had given the prize to Achilles: Achilles is the bravest, Nestor the wisest, and Odysseus the "varied" (*polutropos*, which also signifies "full of tricks"). Nothing new in this classification, and yet Socrates is not convinced. Note that he chose to remain silent during Hippias' peroration and that he does not speak until the public leaves and only a few faithful remain. The stage is set for an exploration of falsity and lying.

The strategy of Socrates, who, as in the *Protagoras*, has the gift of profoundly irritating sophists, is to advance slowly with short questions and answers in order to avoid large passages of rhetoric learned by heart.

Little by little, he leads Hippias to clarify his moralizing evaluation of the characters of the two principal heroes, Achilles and Odysseus. Just as he prefers the *Iliad* to the *Odyssey*, Hippias prefers Achilles to Odysseus. Achilles is fiery where Odysseus is cunning, Achilles is simple where Odysseus is multiple, Achilles is honest where Odysseus is false. Once Socrates has succeeded in bringing Hippias to say that Achilles represents the true and Odysseus the false, he is in his element. Odysseus, *polutropos te kai pseudos*, varied and a liar, is referred to the falsity of the multiple. As Achilles himself declares upon meeting Odysseus, he says something other than he thinks, and is thus a detestable liar: "For I hate him like the gates of death who thinks one thing and says another. But I will speak that which shall be accomplished."[34] In a rather unexpected way, Socrates proceeds to try to rehabilitate the truth of the multiple against this double objection, literary (from Achilles himself) and philosophical (from Hippias).

Socrates begins by proving that saying false is possible only to he who knows the true. A liar is he who is at the same time "powerful and prudent and knowing and wise in those things about which they are false."[35] He who is powerful (*dunatos*) is thus opposed to he who is lacking in this knowledge and this power. A man who makes a mistake in addition is certainly not on the same epistemological level as a man who deliberately gives a false result. In order to lie, it is thus necessary to know the true and the false at the same time. Socrates multiplies examples that he takes from mathematics, geometry, and then astronomy. Flattering Hippias, who boasts of encyclopedic knowledge and who glorifies himself for having invented a mnemonic system, he makes him recognize that, in these cases, in order to speak a falsehood by lying, it is necessary to have a certain knowledge at one's disposal.

Returning next to Homer, Socrates condenses his argument in a nearly paradoxical form, all the while reproaching Hippias for having forgotten his own mnemonic technique. In fact, he does not see a major difference between Odysseus and Achilles when it comes to lying. He explains himself: nowhere in the *Iliad* is Odysseus found to be telling lies, while Achilles contradicts himself on numerous occasions. Achilles announces twice that he is going to return home and leave the battlefield with his army, but he is not seen making any preparations for departure;

he thus says one thing and does another. It is at this moment that Hippias stops Socrates and explains—not without reason—that it is here that the dramatic interest of the poem resides. Achilles' reversals are dictated by his reactions to changes in the situation, like the reverses suffered by the Greek army before Troy. Again, on this point Socrates disagrees: Achilles says different things to Ajax and to Odysseus; sometimes he retires from combat, sometimes he wants to put the Trojan camp to fire and sword. By contrast, Hippias assures him that it is his "innocence" that makes him act in a contradictory manner, while on every occasion Odysseus lies deliberately. Next, Socrates returns to the point on which the two came to agree: he who lies knowingly is better than he who makes a mistake involuntarily. Hippias refuses to follow this principle when justice is at issue; laws punish more severely the person who commits an offense in full knowledge of the facts than the person who is able to protest his ignorance; thus, voluntary justice is harsher than involuntary justice. A contemporary reader would generally be in agreement with Hippias on this point. Even if it can be admitted that lying assumes a surplus of knowledge in relation to ignorance, in the ethical domain, by contrast, an advantage is granted to the error committed in good faith, a frequent technique of the juridical machine in reducing punishment.

Socrates persists and puts forth the opposite argument: ". . . for I maintain that those who hurt or injure mankind, and speak falsely and deceive, and err voluntarily, are better far than those who do wrong involuntarily."[36] As if becoming conscious of the enormity of what he is proposing, Socrates seems to nuance the scope of his assertion, and adds: "Sometimes, however, I am of the opposite opinion, for I am all abroad in my ideas about this matter . . ."[37] Can Hippias cure him of his painful incertitude, which he compares to a sickness or an attack of fever? Socrates is pushed forward by the logic of the dialogue, which forces him to conclude the superiority of the voluntary miscreant over the involuntary one. However, in his mocking reference to the knowledge of Hippias, he had compared the latter to Odysseus![38] As for Hippias, he is not far from refusing to continue this insane dialogue, obviously overwhelmed by Socrates' paradoxes: ". . . but then, Eudicus, Socrates is always troublesome in an argument, and appears to be dishonest."[39] This tactic reappears among the moralizing reasoners: if you are not on their side, you are on the side of evil.

The discussion picks up on the respective merits of the voluntary and involuntary. If one examines a runner, a wrestler, bodily exercise, the voice, physical infirmities such as lameness, and instruments and objects, examples multiply that confirm that the possibility of erring voluntarily is better than that of erring without wanting to. Arts such as equestrianism, archery, and medicine, the most diverse fields of knowledge, are more assured when he who possesses this knowledge is capable of doing well and doing badly all together. Conceding this, Hippias refuses to cede on the level of justice: there, and only there, it is better to err through ignorance. Socrates tries hard to position justice as a type of knowledge, but he cannot convince the sophist. He summarizes his argument, which has become practically absurd: "Then the good man will do wrong voluntarily, and the bad man involuntarily, if the good man is he who has the good soul?"[40] This last *if* carries important consequences. If the good man possesses a good soul, it is certain that he will refuse to do evil in knowledge of the facts, which heralds surreptitiously the Socratic maxim that will be reiterated in the late dialogues: no one is voluntarily bad.

However, Socrates does not succeed in convincing himself. His argument seems to possess an inevitable logic, but it does not gain his approval. When Hippias cannot admit that he who sins voluntarily is a good man, Socrates adds: "Nor can I agree with myself, Hippias, and yet that seems to be the conclusion which, as far as we can see at present, must follow from our argument."[41] The dialogue thus ends in an aporia, a theoretical impasse that will be lived in the form of an irreducible incertitude, and the last word, *plane*, marks a hesitation, a logical doubt close to the error that nothing can efface, not even the official knowledge of those who make a profession of being wise men like sophists: ". . . but if you wise men also wander, and we cannot come to you and rest from our wandering, the matter begins to be serious both to us and to you."[42] The final note is that of an endless wandering that cannot choose between the epic mode (are we so far from the voyages of Odysseus in search of his homeland?) and the skeptical mode (if error haunts errantry, there is no recourse to be taken in any science, and thus all truth is lost). The epic of doubt on which Socrates' argument ends keeps its paradoxical and scandalous edge, and he will at least have demonstrated to the sophist that it was not possible to establish a theory of justice or injustice based on convincing formal reasoning. Socrates seems propelled forward by a logic that he develops with brio, but of which he himself cannot believe all of the consequences, and, like Stephen Dedalus developing his interpretation of Shakespeare, he cannot himself believe in his own theory.[43]

What remains of this dialogue for the reader, despite its abrupt end, is the repeated praise of Odysseus against Hippias' objections. But which Odysseus? Socrates seems to be at his most shocking when he sings the praises of the bad man who is consciously bad: a sadist who acts in the name of his convictions is better than a blind torturer, obeying orders and who thus does not know what he is doing. This would justify Lacan's idea according to which Plato—especially the young Plato who is beginning his writing career after Socrates, the real Socrates, persuaded him to burn his drafts of tragedies and consecrate himself entirely to philosophy—is a distant cousin to the Marquis de Sade: "Plato is a master, a real one, a master at the time when the city-state is decaying, carried away by the democratic barrage that is prelude to the great imperial confluences—a sort of funnier de Sade."[44] In fact, certain passages in the *Lesser Hippias* appear still more shocking to the modern mode of thought than the proto-Nazi drift of the political reveries on blood and land in the *Republic*.

Troy *Doesn't Lie, It's Homer Who Errs*

This problematic is well illustrated by *Troy*, a sometimes ridiculous and barely credible film, often badly acted, another Hollywood delusion that takes the most curious liberties with the Homeric legend. Brad Pitt, an athlete with a bull neck whose American accent is grating compared to the British accents of most of the actors, knows how to prove that he is one of our boys and demonstrates that all-out war is only an extension of bodybuilding. He incarnates the objective lie of the Greeks, while his opponent, more than the Trojans, is Agamemnon. Agamemnon sets a total deception in motion, at the same time as a will to power that does not shrink before any means to satisfy his aims, and to objective victory of the imperialist forces. Agamemnon exposes plainly the imperialist ambitions of the Greeks: the war against Troy offers it the opportunity to be reunited under a single command, and the promise of booty cements their cohesion. The cuckolding of Menelaos and the impertinence of Paris can be forgotten, simple pretexts for the military conquest that will open new markets. As for Odysseus, it is indeed he who speaks the truth in the film, even when he takes recourse in trickery: trickery is one of the weapons in the struggles to the death of warring groups, and, as Paul Faure says, he is often called *polymächanos*, the industrious, the guileful, "because he knew how to take cities by indirect means, by disguise, spying, diversions, simulated flight, false piety."[45] Thus, the famous Trojan horse, epic magnification of an old trick of war by which provisions are brought to famished victims to convince them that combat has ended and to open their gates.[46] This trick is, however, the feat of a realist who wants to end as quickly as possible a war that is dragging on. In the film *Troy*, in one of those added passages that is sublimely anachronistic, Odysseus explains to Achilles that he is indeed the servant of a king at the same time as being a king himself, for he must serve the "king of kings" if he wants to remain king. To repay this frankness, Achilles tells him that he is the only man in the Greek camp whom he respects. In contrast with an Achilles who is tormented, indecisive, caught between a contradictory ethics, politics, and even sexuality, Odysseus displays no romanticism, no desire for death—he just wants to extricate himself from the brutal game of conquest while keeping his wits about him. Consummate diplomat and champion of *poliorcetics* (the technique of besieging cities), he heralds the Italian Renaissance in reuniting the qualities of a Leonardo da Vinci and of a Machiavelli.

Troy, Hollywood *peplos* pumped up on steroids, offers us, as much as *True Lies*, an oblique commentary on American war, whether it unfolds in the Gulf, in Afghanistan, or in Iraq. The armada of Greek ships evokes D-Day, but also the squadron of US fleets in the Gulf. It is not by chance that the real bad guy in the film is Agamemnon, while all of the Trojan leaders are noble and virtuous. Even if it is an absurd extrapolation to show Briseis stabbing the king of kings in vanquished Troy on a hunt for carnage, there is a sort of poetic justice in seeing the priestess of Apollo, captured, raped, and exchanged between princes, striking back and killing the dictator at the moment when he thinks he has achieved his aim. Here, at least, is a character who will not appear in *Troy II: The Return*! This imperfect film succeeds in animating itself with an epic breath when it depicts the struggle to the death between an empire in the process of constituting itself thanks to the collapse of cities and small kingdoms (which will only take definitive form in the Achillean dreams of the kings of Macedonia, Philip, and Alexander the Great), and another regrouping that prefigures a more democratic globalization implying the fusion, the intermixing, of the "multitudes"—multitudes that should be able to be mobilized without forcing them to believe that they are all born of the same land or of a sole mother. Otherwise, the only goddess depicted in *Troy* is Thetis, the mother of Achilles, who forewarns him of his heroic yet brief destiny. Finally, if, as Plato reiterates, the truth comes from the mouth of Odysseus and not Achilles, this film proves that a fiction swarming in anachronisms (don't we see the heroes of a purely oral culture predict that soon their history will be written down?) and distortions can deliver a sort of truth.

All the same, it is Plato, in insisting against sophists like Hippias and Protagoras on the idea that there exists an ultimate truth, locatable in the world of essences and mathematical idealities, who justifies lying most convincingly, while lying horrifies Hippias who nonetheless never stops "lying." And Nietzsche, who proclaims everywhere that he is on the side of the lie, begins to track down all the illusions that remain to us and trace them to their source, the belief in God founded upon grammar (our belief in the possibility of essentializing categories such as the subject, verb, and compliment leads us to the belief in God or in substance, as is to be observed in Descartes); it is through a critical genealogy that he intends to relentlessly denounce them. Thus, the end of the second *Untimely Meditations* already cited opposes a necessary and redemptive truth to the "white lie" proposed by Plato in the form of the autochthonic myth of origins of the children of the republic. Such a "necessary truth" will be critical and necessarily demystifiying: as the German culture suffers from a sickness which is the sickness of history, it is necessary to make young students conscious of a work they need to accomplish themselves in order to rid themselves of the lying vestiges of their education. This education has become a second nature that it is necessary to abolish. Nietzsche cites a Spanish proverb, *Defienda me Dios de my*:[47] "God, protect me from myself"—that is to say, from the nature already educated into me. This truth must be drunk, drop by drop, like a bitter remedy. The necessary truth will thus struggle against history, against culture, all the while finding in the Greeks a model of the creative faculty that was able to forge a new synthesis from the most diverse influences—Semitic, Babylonian, Lydian, Egyptian—because the fundamental value of life had never been excluded from it.

Life will thus be the source of all truth—but how to accede to this knowledge? Isn't it frequently the intuitive object of the poets? In *Thus Spoke Zarathustra*, a powerful theme is that of the excess of poetic figures who lose sight of the truth. The chapter "Upon the Blessed Isles" affirms: "All the permanent—that is only a parable. And the poets lie too much."[48] Why, then, do the poets lie? They lie in the sense that they eternalize everything, and no longer know how to sing the ephemeral. Nietzsche, who sees himself as a poet and a philosopher, endeavors to lie as little as possible, but, like Socrates, is pushed forward by the logic of his argument into lyrical parables. According to the critical analysis of Heidegger, the objective irony is that Nietzsche falls again into the most ancient myths of the metaphysical West when he places confidence in the will to liberate himself from myths and lies. "But thus my creative will, my destiny, wills it. Or, to say it more honestly: this very destiny—my will wills."[49] Or, more literally: "Such a destiny, it is my will that wanted it."[50]

If it is necessary to speak more openly to confess this truth, it is because the poets lie too much—for the philosophers. Nietzsche follows here the path opened by Heraclitus, who had declared that Homer deserved "to be cast out of the contests and beaten with the staff."[51] For Heraclitus, as for Nietzsche, the child is "innocence and forgetting," just like the formula of Zarathustra: the child lives in the present because he still has not entered into the world of history and of official lies. By contrast, Homer, who was said to be so wise, was no more valuable to the intelligence than a couple of filthy scoundrels. This is what is revealed to us by the famous anecdote concerning the death of Homer, worn out and discouraged, unable to understand a childish enigma: "People are thoroughly beguiled regarding the knowledge of manifest things, as was

Homer, who was wisest of all the Greeks. For he was beguiled by boys killing lice, who said: 'what we saw and seized, those we leave behind, what we neither saw nor seized, those we bring.'"[52]

This scene opposes the blindness traditionally reserved to *Homeros*, the blind bard with the generic name, to the partial vision of the children who only capture the lice that they see but, in contrast to the fish that one fishes for, carry away with them the ones they don't see. We have observed how it suffices to have the truth in mind to be able to lie—that is what Socrates brilliantly demonstrates, persuading himself and convincing Hippias that a dual lying knowledge is better than a simple knowledge that is erroneous or marked by lacunae. Can it be said that the truth that we carry in spite of ourselves is like the lice the children bring back to their homes? This Freudian idea—that the truth is that part of the invisible that remains in us despite our efforts to see it or eliminate it—would not be inadmissible to a Greek, at least not to a philosopher such as Aristotle, who believed that some lice were born

from simple contact between the skin and sweat. What is more, he believed that this parthenogenesis of lice was more visible on the curly heads of our children.[53] If the truth is sordid, it is because it multiplies itself, because it returns stubbornly despite our efforts to hide it; and if the lie does not come to reduce it, it will be some kind of game, enigma, or riddle that is able to purify it.

Heraclitus presents children who pose a riddle resembling a Freudian *Witz*. They mock the old poet and playfully summon him to tell them what he thinks of the critique—of the act which consists in choosing what is to be thrown away or kept with oneself, even on the body and often in spite of oneself. Here, the lie is thus only a sport, a game, a "language game," to put it in Wittgenstein's words, which, however, does not share the confidence accorded by Nietzsche to the child. The disabused remark in the *Investigations* is well-known: "Are we perhaps over-hasty in our assumption that the smile of an unweaned infant is not a pretence?—And on what experience is our assumption based? (Lying is a language-game that needs to be learned like any other one.)"[54]

Heraclitus wanted himself to be the Obscure—not to deceive, but only to hide in revealing, acting thereby like Nature, of which the essence was, according to him, to hide and to hide herself. Must it be concluded that the poets no longer lie if they depict liars? If that is the truth, then there is no longer any doubt: the true hero of Homer (if he ever existed, if he wrote both the *Iliad* and the *Odyssey*), is Odysseus, the lying hero, the astute strategist, the industrious, the cunning man with a thousand tricks, *polutropos*. Sometimes, certainly, it seems he goes a little too far—doesn't he continue to lie to his father just after he massacred the suitors and became reunited with Penelope? In Canto XXIV, visiting his father whom he has not seen for twenty years, he decides to "test" Laertes with "mocking remarks," and for nearly forty lines invents a story about Odysseus in which he had supposedly given him four women as a gift. When Laertes twice enjoins him to be sincere and tell the whole truth, Odysseus recounts that he is called Eperitos, that he comes from Alybas, and has not seen Odysseus for five years! It is necessary for his father to express the most intense mourning by covering his head with ashes for Odysseus to finally be moved and reveal his identity. It is Laertes who demands visible signs in order to avoid deception. The Hellenist Faure emphasized that the name Odysseus was typically Cretan. If it is admitted, following Faure, that Odysseus is Cretan, this heralds all the paradoxes to come: Odysseus is at the same time an impenitent liar and a tragic hero condemned to go forward. Is there a big difference between the Socrates, always skeptical and always buffeted by unsupportable arguments in the *Lesser Hippias*, and the lying Odysseus, heroic and condemned for his *hybris* in Dante's poem? If he sinks in this final shipwreck, this epic radicalization of the *plane* of Socrates, we will nonetheless not be able to totally discredit the enthusiastic words he uses to rally his former companions one last time toward adventures condemned in advance:

> Consider well the seed that gave you birth:
> you were not made to live your lives as brutes,
> but to be followers of worth and knowledge.[55]

Notes

1. Claude Zidi et al., *True Lies*, directed by James Cameron (July 15, 1994) (Los Angeles: Twentieth Century Fox, 1994). Based on Zidi's French film *La Totale!*, Paris: AMLF, 1991.
2. Barbès is a neighborhood in northern Paris known for its high proportion of immigrants.
3. Stanley Cavell, *Pursuits of Happiness* (Cambridge, MA: Harvard University Press, 1981).

4. Friedrich Nietzsche, "On Truth and Lying in an Extra-Moral Sense," in *Friedrich Nietzsche on Rhetoric and Language*, ed. and trans. Sander L. Gilman, Carole Blair, and David J. Parent (New York: Oxford University Press, 1989), 246–57.

5. Under the title Nietzsche, *Introduction théorétique sur la Vérité et le Mensonge au sens extra-moral*, in *Le Livre du philosophe: études théoriques*, trans. Angèle. Kremer-Marietti (Paris: Aubier-Flammarion, 1969; repr. 1991), 117–33. English translation: *Unpublished Writings from the Period of Unfashionable Observations. Complete Works*, trans. Richard T. Gray (Stanford, CA: Stanford University Press, 1995).

6. Nietzsche, *Unpublished Writings. Complete Works*, 189. See also Nietzsche, *Werke; Kritische Gesamtausgabe*, ed. Giorgio Colli and Mazzino Montinari (Berlin: Walter de Gruyter, 1978), sec. 3, vol. 4, 230. A useful selection is found in Nietzsche, *Über Wahrheit und Lüge*, ed. Steffen Dietzsche (Frankfurt: Insel, 2000), 33.

7. Nietzsche, *Unpublished Writings. Complete Works*, 198; and Nietzsche, *Werke*, sec. 3. vol. 4, 240.

8. I am summarizing here the theories of Giambattista Vico in *La Science nouvelle*, trans. Ariel Doubine, ed. Fausto Nicolini (Paris: Nagel, 1953), 118 on the primitive giants, 124–25 on Zeus and thunder, and also 254–55 on civilization. Note that Vico is one of the rare thinkers to have attempted to understand the historical genealogy of the myth of Platonic "autochthony," and thus the beliefs that make certain human groups consider themselves to be "the most ancient" or "aboriginal," and thus, inevitably, "sons of the Earth" (118).

9. Nietzsche, *On Truth and Lying*, 255.

10. Lucian of Samosata, *Works*, trans. A. M. Harmon, Loeb Classical Library (Cambridge, MA: Harvard University Press, 1995), vol. 3, 323.

11. Stories about phantoms and all sorts of apparitions are the modern version of these old wives' tales attacked by Lucian. I have written about them in connection with philosophers such as Spinoza and Schopenhauer in *La Pénultième est morte* (Seyssel: Champ Vallon, 1993). English translation: *The Ghosts of Modernity* (Gainesville: University Press of Florida, 1996).

12. Nietzsche, *On Truth and Lying*, 254.

13. Friedrich Nietzsche, *The Birth of Tragedy*, ed. Raymond Geuss and Ronald Speirs, trans. Speirs (Cambridge: Cambridge University Press, 1999), 4–5.

14. Ibid., 5.

15. Friedrich Nietzsche, *Dernières lettres*, trans. Catherine Perret (Paris: Rivages, 1989).

16. Max Scheler, *Ressentiment*, ed. Lewis A. Coser, trans. William W. Holdheim (New York: Free Press of Glencoe, 1961), 77–78. See also Max Scheler, "Das Ressentiment im Aufbau der Moralen," in *Gesammelte Werke* (Berne: Francke Verlag, 1955), vol. 3, 67.

17. Nietzsche, *Unpublished Writings. Complete Works*, 190, and Nietzsche, *Werke*, sec. 3. vol. 4, 231.

18. Nietzsche, *Unpublished Writings. Complete Works*, 192.

19. Friedrich Nietzsche, *Untimely Meditations*, ed. Daniel Breazeale, trans. R. J. Hollingdale (Cambridge: Cambridge University Press, 1997), 118. See also Nietzsche, *Unzeitgemäße Betrachtungen* (Munich: Goldmann, 1964), 138.

20. Plato, *Republic*, 2, 382c, in *The Collected Dialogues*, ed. Edith Hamilton and Huntington Cairns (Princeton, NJ: Princeton University Press, 1996), 629.

21. Ibid., 2, 381d, 628.

22. Ibid., 2, 381e–382a, 628–29.

23. Ibid., 3, 381b, 634.

24. Ibid., 3, 415de, 659.

25. Ibid., 659.

26. Ibid., 660.

27. Ibid., 659.

28. See Marc Shell, *Children of the Earth* (New York: Oxford University Press, 1993).

29. Plato, *Republic*, 3, 415de, in *The Collected Dialogues*, 659.

30. Plato, *Laws*, 663d, in *The Collected Dialogues*, 1260.

31. Plato, *Statesman*, 271ab, in *The Collected Dialogues*, 1036.

32. In French, *répétition* also has the strictly theatrical meaning of "rehearsal."

33. Friedrich Nietzsche, *Human, All Too Human*, trans. R. J. Hollingdale (Cambridge: Cambridge University Press, 1986), 40.

34. As Achilles says in the *Iliad*, canto 9, v. 311–12, cited by Plato, *Lesser Hippias*, in *The Collected Dialogues*, 202.

35. Ibid., 203.

36. Ibid., 209.

37. Ibid.

38. Ibid., 208.

39. Ibid., 210.

40. Ibid., 214.

41. Ibid.

42. Ibid.

43. See the conclusion of the discussion in James Joyce's *Ulysses* where Stephen Dedalus attempts to demonstrate that Shakespeare is his own grandfather and thus the father of all writers. Eglinton asks him: "Do you believe your own theory?—No, said Stephen promptly."

44. Jacques Lacan, *Le Seminaire VIII: Le Transfert* (Paris: Seuil, 1991), 106.

45. Paul Faure, *Ulysse le Crétois : XIIIe siècle avant J.-C.* (Paris: Fayard, 1980), 167.

46. "This concerns a strategy repeated many times throughout history, from the taking of Jaffa by the Egyptian Thoutii, the general of Pharaoh Thutmose III, around 1460 B.C., to the seizing of the fortresses of Fougeray and de Mantes by the Briton Bertrand Du Guesclin, two thousand eight hundred years later" (ibid., 169).

47. Nietzsche, *Untimely Meditations*, 119. See also Nietzsche, *Unzeitgemässe Betrachtungen*, 139.

48. Friedrich Nietzsche, *Thus Spoke Zarathustra*, trans. Walter Kaufmann (New York: Penguin, 1978), 86.

49. Ibid., 87. See also Martin Heidegger, *Nietzsche* (Pfullingen: Günther Neske, 1961).

50. Friedrich Nietzsche, *Also sprach Zarathustra* (Münich: Goldmann Verlag, s.d.), 66.

51. Heraclitus, "Fragment XLII," in *Fragments: Translations and Analysis*, by Dennis Sweet (Lanham, MD: University Press of America, 1995), 19.

52. Ibid., 25.

53. Aristotle, *Historia animalium*, trans. A. L. Peck, Loeb Classical Library (Cambridge, MA: Harvard University Press, 1970), vol. 2, bk. 5, 549b5–13 and 556b25. I examine the way Joyce treats the problematic of autogenesis or the parthenogenesis of lice in theoretical (see notes on Aristotle and on aesthetics) and fictional (*Portrait of the Artist as a Young Man*) texts, in *James Joyce and the Politics of Egoism* (Cambridge: Cambridge University Press, 2001), 85–106.

54. Ludwig Wittgenstein, *Philosophical Investigations*, trans. G. E. M. Anscombe, 3rd ed. (New York: Macmillan, 1970), 249, 90e. See also Wittgenstein, *Philosophische Untersuchungen* (Frankfurt: Suhrkamp, 1967), 115.

55. Dante, *Inferno*, trans. Allen Mandelbaum (New York: Bantam, 1982), 245.

Roald Hoffmann

Honesty to the Singular Object

WHAT ETHICAL LESSONS, IF ANY, MIGHT EMERGE FROM NORMATIVE science? My nonlinear path to a response begins with a look at storytelling in science. The moral implications of narrative will then quickly take me to an exploration of the ethical considerations that science might (or might not) call forth.

Stories

Science tells some rollickin' good stories. So why are scientists so unappreciative of the necessity of storytelling for the success of their own enterprise? Why do they beatify Ockham's razor rather than the rococo inventiveness of their hypotheses?

Because they are afraid of "just so" stories. The Kiplingesque allusion points to one of science's historical antipathies—to the teleological. Countered by a human proclivity for exactly that, the teleological, in the retelling of scientific stories. Could there be something else astir, for instance, a suspicion of the particularity of language, when scientists are ideologically committed to infinitely paraphraseable universals?

Consider first the marvelous stories that emerge out of science. So many to choose from—the epics of continental drift, or the way one iron atom in hemoglobin communicates with another. Or amusing ones, such as how the amount of vanilla claimed to be natural in French ice cream exceeds by a factor of ten the quantity of beans shipped from Madagascar. Which led to a cat-and-mouse game between the forgers of vanillin (the flavor principle here) and the scientific detectives who learned to distinguish between the natural and synthetic form of one and the same molecule.[1] Or take a triumph of molecular biology, the working out of the chemistry and function of the ribosome. In figure 1 is a schematic illustration—not an atom in sight in this low-resolution representation—of this biomolecular "smart" factory. It is a complex of about eighty proteins and a few RNA molecules that takes a strand of messenger RNA (complementary to DNA) and initiates a process of linking, according to the RNA instructions. The amino acids are shuttled to the ribosome by a transfer RNA, these attached to an already formed piece of the enzyme. And proofreading the enzyme coming off, at a rate of twenty amino acids a second.[2]

Shall I compare thee to a Rube Goldberg machine (figure 2)?[3] (In England it would be Heath Robinson.) And is there a gaping trap in

Figure 1. Image of a subunit of a ribosome, by Ada E. Yonath.

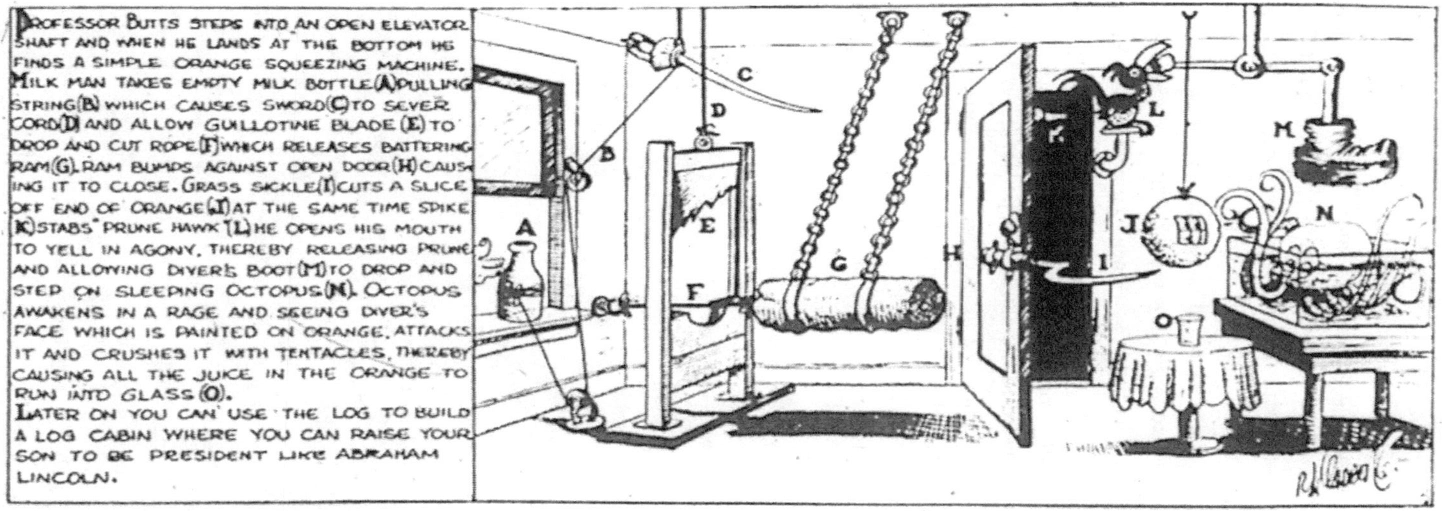

Figure 2. Artwork © TM Rube Goldberg Inc., used by permission.

this simplistic mechanism of mechanistic visions? Oh, yes. The way we envisage the ribosome is mechanical, linear, and . . . ephemeral. The representation, thrilling as it is, is transitory. Yet—and this is what some critics of scientific knowledge miss—this most unfaithful representation doesn't hinder us from designing real, functional antibiotics that throw a wrench into the workings of microbial ribosomes.

The ribosome story allows me to shift to something much more interesting. This is the utility—nay, the necessity—of storytelling for practicing science.

Why should storytelling be essential for science? Well, every time the simple is proffered, human beings fall for it. So admiration for the symmetrical molecules, exemplified by the ones shown in figure 3, or for a simple mechanism of a chemical reaction, the aesthetic imperative in physics (if an equation is beautiful, it must be right) seems . . . natural. And, I would add, related to our falling for political ads, of any persuasion.

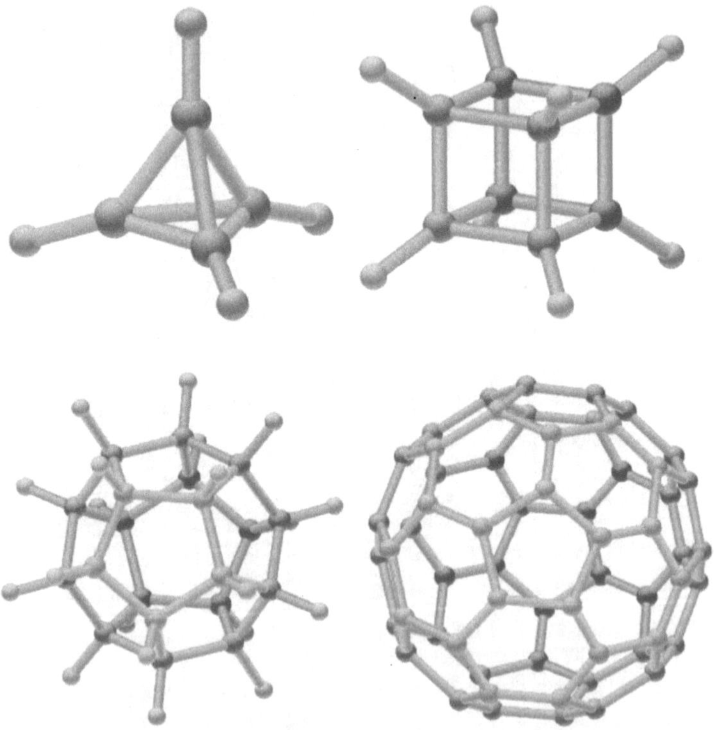

Figure 3. Symmetrical molecules.

But what if honest investigation of the real world reveals complexity, bound to be discovered in any biological or cultural entity that has been subject to inherently complexifying evolution? Even in a molecule. Take a look at hemoglobin, the oxygen carrier in our blood (figure 4). This is already a much-simplified representation, omitting the vast majority of the more than 9,000 atoms in this $C_{2954}H_{4516}N_{780}O_{806}S_{12}Fe_4$ molecule. Where, and how, does one then find pleasure in such contorted complexity?

By telling a story, weaving one: the story of hemoglobin, of its four subunits, of the α-helices curled in it, of the way an iron atom held in the middle of a platelet-shaped heme molecule binds the oxygen, and in doing so pulls on a histidine group, . . . Storytelling seems to be ingrained in our psyche. I would claim that with our gift of spoken and written language, this is the way we wrest pleasure, psychologically, from a messy world. Scientists are no exception. We tell stories because they first satisfy, then keep us going. Stories "domesticate unexpectedness," to use Jerome Bruner's phrase.[4]

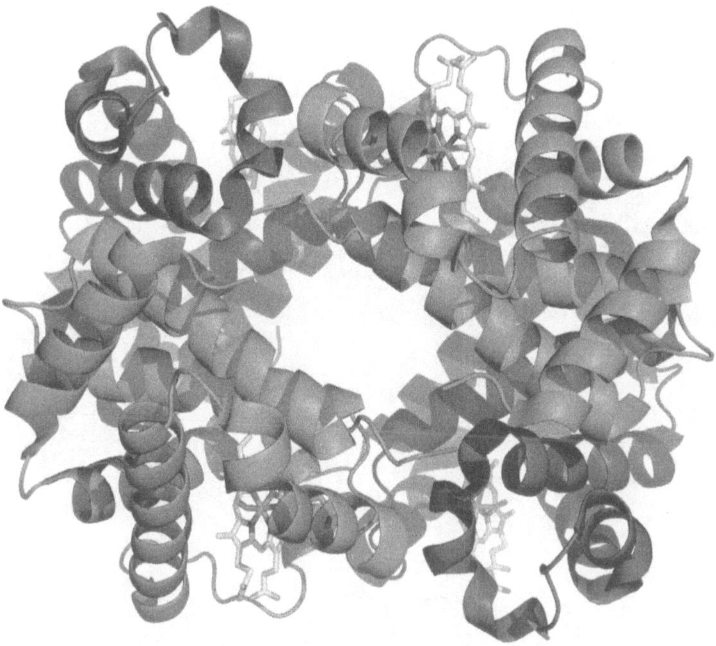

Figure 4. A schematic "ribbon" drawing of the structure of hemoglobin.

A Short Story of the Natural World

Let me tell one such constructive story. Insects are the greatest chemists. They use simple chemicals in communication, mating, defense, and predation. In 1966 R. S. Berger identified the main sex pheromone of the cabbage looper moth, *Trichoplusia ni (Noctuidae)*, shown in figure 5 in its caterpillar stage, as "(Z)-7-dodecenyl acetate," a simple molecule related to some fatty acids in all living things. This pheromone is also illustrated in figure 5.

Those were the halcyon days of early pheromone chemistry; everyone was happy with one molecule (as they were with one gene for each trait). Thirteen years later, L. B. Bjostad et al. identified a second component, important especially in close-range courtship behavior. Then the same group began to think through the biosynthetic relations between these two components and other molecules observed in the pheromone gland. Obviously, enzymes that do various transformations—shorten molecules, remove hydrogens, add various atoms, all the wondrous machinery of the living—are at work. I show in figure 6 a complex graph from one of the

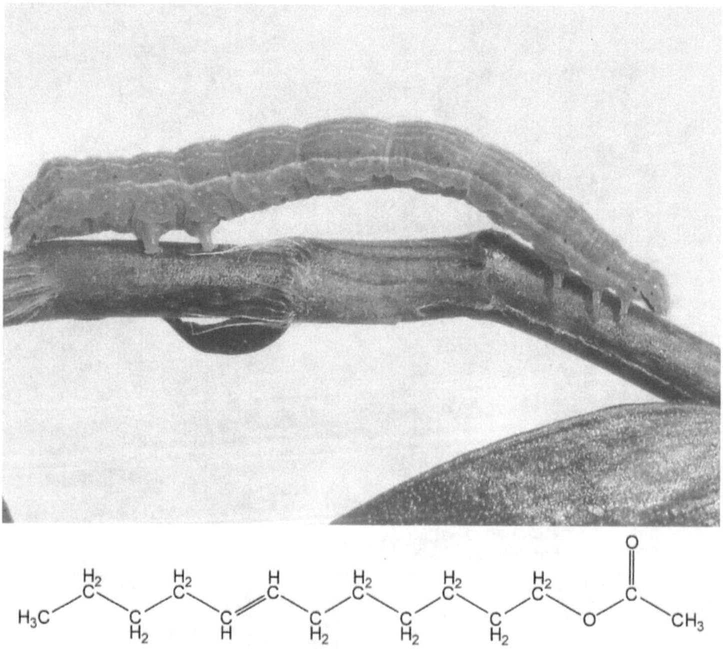

Figure 5. Cabbage looper moth. Photograph by Mark O'Brian.

Bjostad et al. papers, indicating the biochemical relations between the various kinds of fatty acids in the moth.[5] Here's the story, a biochemical story that moved the authors, which they confide to us. A blend of six components, suggested by their analysis of the way the molecule first thought to constitute the pheromone is made by the insect, elicited complete courting flights against a stiff breeze in a wind tunnel. Clearly one needs six for sex. And would a human master perfumer be surprised?

The story is told with sufficient verve in the Bjostad, Linn, Du, and Roelofs paper that even I, an outsider to the field, am pulled in by it. More

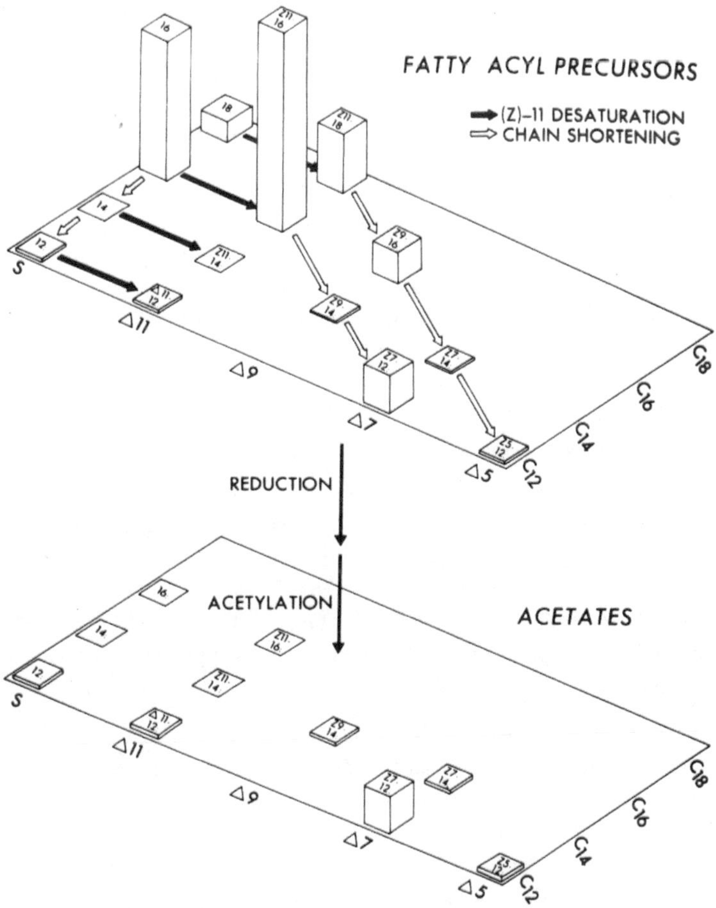

Figure 6. Biosynthetic clues to pheromone mixture. Reproduced from reference 5 with permission of Springer Science & Business Media.

than just an analysis of pheromone glands, the biochemical relations are clever. I am intrigued by their tale and begin to think of its sequel—how do the females evolve that blend? How do the males evolve the receptors to it? Thomas C. Baker, and his co-workers, professors of entemology ahd chemical ecology, at Iowa State University have actually located separate compartments for the six components (and one so-called antagonist, a molecule that acts to negate the physiological reaction to the pheromone) near where the male antenna input is first processed. Extending the story is life-enhancing. And not just in the *Thousand and One Nights.*

John Polanyi has recently described the close relationship between science and storytelling:

> *Scientia* is knowledge. It is only in the popular mind that it is equated with facts. This is, of course, flattering, since facts are incontrovertible. But it is also demeaning, since facts are meaningless. They contain no narrative. Science, by contrast, is story-telling. That is evident in the way we use our primary scientific instrument, the eye. The eye searches for shapes. It searches for a beginning, a middle, and an end.[6]

The power of stories may indeed exceed that of facts. As Walter Benjamin has written:

> The value of information does not survive the moment when it was new. It lives only at that moment; it has to surrender to it completely and explain itself to it without losing any time. A story is different. It does not expend itself. It preserves and concentrates its strength and is capable of releasing it even after a long time.[7]

In telling the story of scientific discovery, we form a praiseworthy bond with literature and myth, all the other ways that human beings have of telling stories. Yes, there are times when the story has to be told simply, the fire engine sent the shortest route to the fire. But a world without stories is fundamentally inhuman. It is a world where nothing is imagined. Could a chemist be creative in such a world?

Moral Endings

Almost every story has a moral, explicit or not. As Hayden White asks: "When it is a matter of recounting the concourse of real events, what other 'ending' could a given sequence of such events have than a 'moralizing' ending?"[8] Emily Grosholz notes perceptively that "when we hear a story, we evaluate the agents and the action. Maybe this is because of the irrevocability of human action (it only happens once, so it better be good), and intentionality (we always do things for an end or reason). . . . The very choice of beginning and ending confers meaning."[9]

In the endings I have seen, more often in scientific seminars than in fragmented papers, there is a curious mixture of celebration of the human achievement with serendipity co-opted to serve design. And there's a double resacralization of the mundane, of what has just been demystified. In a spiritual process that I find refreshing, awe is expressed at what happens in nature or what human beings (okay, usually the author) can do. And the mysteries yet to be resolved are articulated. With complete faith that they will be resolved.

Ethics Growing Out of Science?

The significance of storytelling in science, and the attendant morality, leads me to think of the potential of the process of science for constructing ethics. Here I follow, half a century later, mathematician Jacob Bronowski's path. My landsman in more ways than one was much less tentative than I will be when he said, "the practice of science compels the practitioner to form for himself a fundamental set of universal values."[10]

Can ethics grow out of science? The very question may seem ludicrous to two communities of scientists: those who really believe that science is ethically neutral, and those who believe that scientists are inherently ethical. So let me first contend with these, as provocatively as I can.

To claim that science is ethically neutral ("I just worry about the technology of cloning; someone else can decide if it's good for people") puts scientists squarely in the company of anti-gun-control activists ("Guns don't kill, people do"). By contrast, I believe, and there is some philosophical tradition that supports this, that in any action by a human being, the instrument of that action (a gun, a molecule synthesized, yes, even a mathematical equation or a poem) must be accompanied by a moral judgment. The judgment is: "Will the use of that instrument by me (or by others) hurt people, or not?" The invention or implementation of a tool without consideration of the consequences of its use is deeply incomplete.

As for the claim that scientists are born with ethics—well, that's just as likely as their being born with aesthetics or logic. That the latter is not true, you learn from reading the "peer review" referees' comments on your paper. We scientists are people who have opted to engage in a remarkable social system for garnering reliable knowledge, that knowledge being of great practical and spiritual value. The critical components of that Western European social invention, science, are (a) normal, curious people, some of whom like mathematics; (b) people not afraid of getting their hands dirty—experimenters; (c) an open system for dissemination of what one finds, and a communal urge to do so; and (d) a method that encompasses frequent dipping back and forth between approximations of reality (gauged by our occasionally misleading senses and our tools) and flights of imaginative fancy in hypothesis formation and theory building.

So . . . is there something in the practice of science that can enhance the ethics brought to it by scientists, or that possibly can engender an ethical outlook? As the above makes clear, I do not come to this because I think scientists are "better" than other people—far from it. Nor do I dare presume that a relatively late social invention, science, could provide a broad rationale for a human quality as fundamental as ethics. (Or is ethics itself a social invention? If so, it is older than science. But not as old as curiosity.)

When goods collide, where do we get our criteria for deciding among them? From the usual sources, like them or not: our socialization at home and in our schools, that is, from our parents and teachers. Perhaps from our genes, though not as much as biologist E. O. Wilson would like us to believe. From churches and religions. From reading—novels are especially strong moral instruments. Not a tad diminished by deconstruction. By the time science enters a young person's moral consciousness, he or she is usually a well-defined moral human being. Yet the web of life has a way of generating new quandaries; one's personal sense of what's right, how to act in difficult times, evolves even as it is moored in the past.

Two of the components I gave of science, publishing and the nervous motion twixt theory and reality, depend on texts, talks, and conversations. These generate narrative. And, even forgetting the moralizing endings, such acts of communication inevitably confront scientists with ethical choices—to be faced, evaded, negotiated. Let me expand on this.

Writing It Down

An important part of the system of science is publication, with the potential of replication. How reproducible scientific findings are (and whether the reality of reproducibility is essential to belief) is a matter of contention.[11] It took several years for public questioning to surface of the all-too-novel measurements, a multitude of them, of Hendrick Schoen in solid state physics.[12]

Could it be that the primary emotional motivation for a scientist who does not falsify a synthesis or measurement is simply fear, rather than the psychoethical drive to report facts honestly? Perhaps, though I find definite positive value in fear in making us behave righteously. To a point. And fear of damnation, big and small, is certainly important in Christian ethics. It may be painful for most of us to see others, never ourselves, "do the right deed for the wrong reason," as T. S. Eliot says in *Murder in the Cathedral*.[13] But I accept the way we are: the "habit of truth," as Bronowski called it, is formed in many ways.

Ethics is like a limb that needs exercise to function. The importance of publication is that it provides exposure to potential testing, time and time again. Fraud in science is ultimately unimportant. There is much prurient interest in it, for sure. With the same origins as our fascination with the sexual misdeeds of our ministers. Priests of the truth have a longer way to fall. But fraud is unimportant because the psychopathology of its perpetrators is such that their fear of being proven wrong is somehow abrogated, and they never forge the dull, only the interesting. Thus, the normal workings of the system ensure that others—out to prove the makers of the startlingly new wrong, not right—will repeat the experiment.

So the system works, but is the individual scientist motivated by loss of reputation if proven deceptive likely to become more ethical? A cynical viewpoint is that he or she will learn to sanitize, embroider, and manicure just enough to get away with what he can, then pile on the hype. A more charitable viewpoint is that we learn that data are not only not to be trusted, but that they are mute and inherently conservative. That a human being must interpret them—yes, tell a story about them. And that it is all right (within a self-correcting system such as science) to risk an imaginative, ornate hypothesis that does an end run around Ockham's razor.

Something salutary takes place in the writing of an experimental part of a scientific paper. I have trouble in picking one of my own to show you, for, sad to say, I'm just a theoretician. But here's a piece of a mixed experiment/theory paper in which I am a co-author, *ergo* in part responsible:

> Crossover Experiments. In a 25-mL reaction flask was placed 0.050g (0.094 mmol) each of $Cp_2{}^*Th(^{12}CH_3)_2$ and $Cp_2{}^*Th(^{13}CH_3)_2$. The vessel was evacuated, and then 10 mL of Et^2O was condensed into the flask at −78 °C. The suspension was stirred at this temperature until all of the material had dissolved and a colorless solution was obtained. The flask was then backfilled with 1 atm of CO and the solution stirred vigorously. After 4 h at −78 °C, the solution was allowed to warm to room temperature whereupon a colorless solid ($[Cp_2{}^*Th(\mu\text{-}O_2C_2\text{-}(CH_3)_2)]_2$) precipitated. Next, 2 mL of degassed 1 M H^2SO^4 was added to the reaction mixture via syringe under a flush of argon. After the resulting suspension was stirred for 15 min, the mixture was centrifuged to remove a colorless, flocculent solid.[14]

Basta! You see a report of what was done, almost in iambic pentameter. Not the average run, but the best that was done, to be sure. It's there, this experimental part of a longer paper, for historical reasons: as evidence that it was done, that it can be done, with details reproducible by anyone (well, maybe). But why give the evidence? Isn't there trust in the community, aren't we all gentlemen? Or were, now that 38 percent of PhDs in chemistry in the United States are women . . .

In citing another's experimental (or theoretical) work, there's a similar wrestling match on. To cite is an act of trust. Which can also be viewed as an act of mistrust, for by citing someone else without questioning the result, one is protected should it be faulty. To say that the mistrust complicit in the statement of conditions of an experiment, or citing someone's work, negates the trust overtly expressed by using the work, and that that's all there is to science, is to miss the fertile, tilled orchard of science—the creation of molecules as well as frameworks of understanding.

This is the essential tension of which Thomas Kuhn wrote, between trusting and not trusting.[15] I think writing an experimental part of a paper, or reading it in someone else's text, not once but many times

(I have written five hundred such, not atypical) is an ethically productive action. One in which both subconsciously and overtly the issues of trust and mistrust are negotiated by chemists. The important word here is *negotiation*: the web of habitual description and citation subconsciously (and explicitly) forces the creator to confront the other. It is an inherently social web, built out of real and imagined interactions with other human beings. In it are the makings of a gift economy. And of empathy.

First-time Narratives

I see two other places where ethics emerges in an interesting way from normative science. The first is the responsibility taught by first-time narratives, first-time representations. I remember for instance, when Fred Hawthorne, professor of chemistry—now at UCLA—came one day in 1961 to Harvard, where I was a graduate student, and told us how he had made $B_{12}H_{12}^{2-}$, a molecule shaped like an icosahedron (see figure 7).[16] Nothing

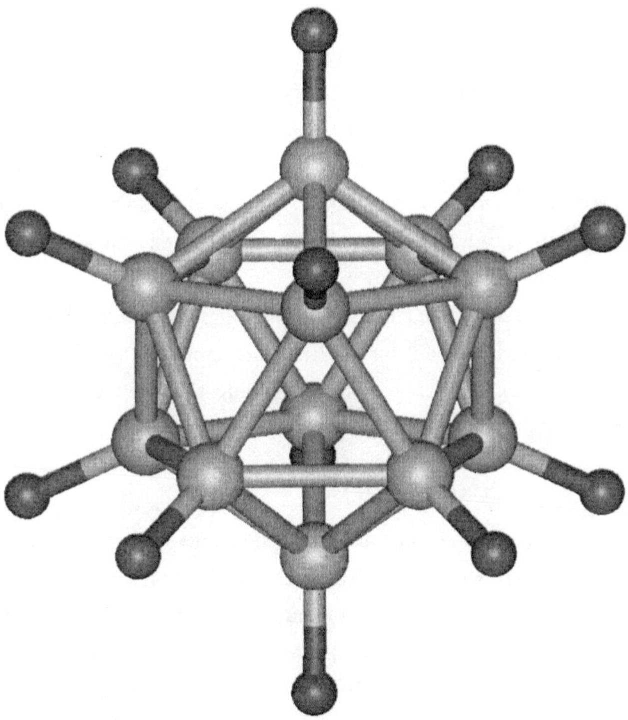

Figure 7. Structure of the $B_{12}H_{12}^{2-}$ ion.

like it had been seen before; he described its properties with evident and appropriate excitement. Hawthorne knew instinctively that there was a story to be told of $B_{12}H_{12}^{2-}$, that it sufficed to tell it straight. *Das Ding an Sich* was indeed beautiful enough; it was sacred even as it came to be in his profane hands. In another day, another time, Fred would have said that it was given to him by the grace of God. In 1961 he called it serendipity.

There was no more question of Hawthorne making up a fib around $B_{12}H_{12}^{2-}$ than of Haydn writing a dissonant section in one of his piano trios.

Representation as Furniture

Much of what we do in science is to represent reality. Those representations, whether in language or not, are murky mirrors. But, as Emily Grosholz says, "representation is also generative: we say more than we know we are saying, and we induce order by our orderings, and good representations, as intelligible things, add to the furniture of the world.

So representations are both more and less than what they represent. By misrepresenting, they also allow us to know, and to create."[17]

Speaking of language and veracity, Oliver Sacks tells an interesting story in a film he made in *The Mind Traveller* series. In Eureka, California, he met a family of deaf Mexican farm laborers. Among the five children, the three older brothers, who did not sign, were suspicious of their younger siblings, who were learning sign language because they would learn to lie, the older ones said.[18]

When you see something for the first time, you don't know what it is. When you describe it for the first time, language will fail you. You grope for meaning. But there is no lie. Would that we were given more such moments!

Honesty to the Singular Object

A second experience is one shared by poets and scientists. Something is seen, felt, then described. Now not for the first time, but for the umpteenth. So love has fled, and it hurts to remember what was good. It has happened to others, though that thought seems not to comfort at all. A poem needs to be written—one is in the Luberon, in winter. One walks out in the morning into the vineyard; it's sad to face that beauty alone. But then there's a grape cluster, like no other grape cluster. It must be described:

RAISINS FOR BEING

They left small bunches
on the vine, green late-
comers; the farmers

knew the day to pick,
sugar rising in the
berries, rain offshore. But

four sunny days broke
the pattern; the vines free
of their luscious burden

filled out the stragglers.
And then I came, just
before pruning,

and walked out in
the morning frost, the sun
clearing the Luberon,

and a thousand droplets,
on a grape cluster,
muscat pavé, told me

that I had a latecomer's
right, to live life out
reflecting, free albeit

tethered, at an angle
to the sun, sweet to you.[19]

I describe, and I am not sad any longer. For a while.

Elsewhere there is a molecule I see in a journal (figure 8).[20] I talk about it to one of my graduate students, Pradeep Gutta. It has at its center a ring, with two tins and two nitrogens in it. But as you see, the environments of the two tins are strikingly different. Why? Could

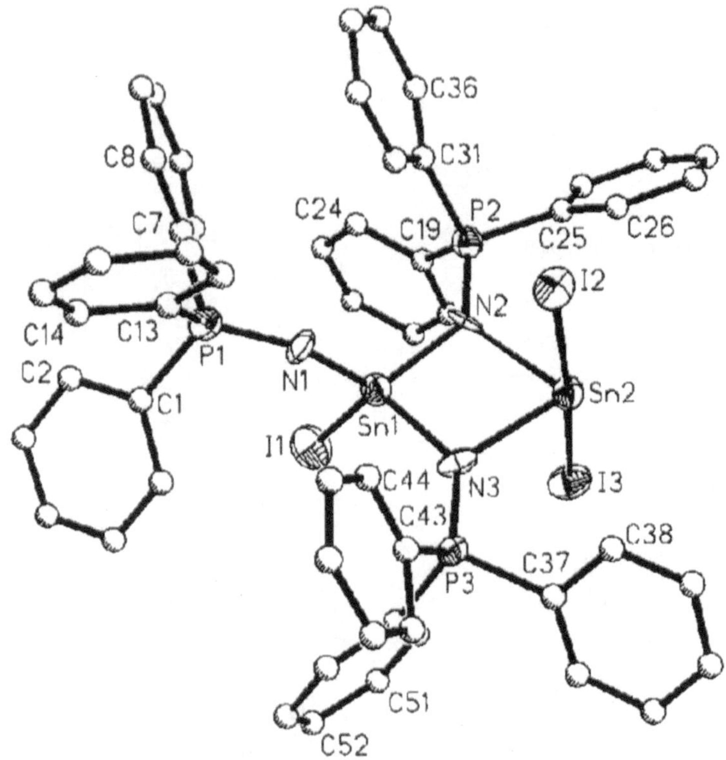

Figure 8. The structure of a molecule made by Chitsaz et al., $[Sn_2I_3(NPPh_3)_3]$

it be because of their different substituents, the chemical shrubbery hanging off the tin atoms? No, for the molecule with all substituents identical is calculated to have exactly the same geometry. "Is a puzzlement," as Yul Brynner said. And could one exchange the environments of the two tins, as shown in figure 9? We calculate the way the elec-

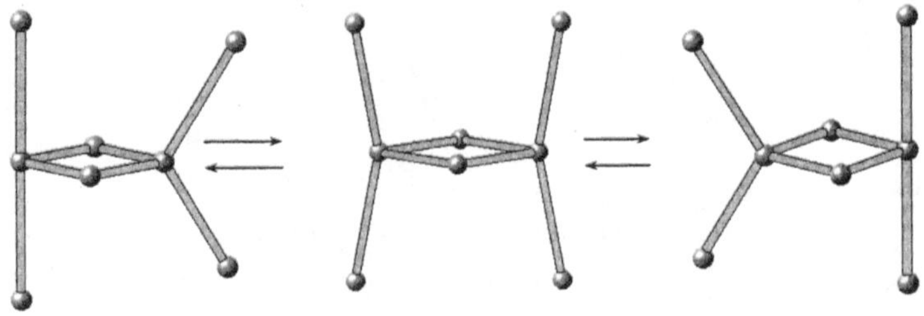

Figure 9. Drawing of the same molecule as in figure 8, pruned to geometrical essentials.

trons move in this molecule, their orbitals, orbits writ large. And we reason out a reason, because . . . that's our *métier*. There's a story to be told. I tell it, as well as I can.

The language I use to tell my story is that of science, which is not the language of poetry, at least not much of the time. There is no premium on ambiguity in science. That a word can mean two things and sound like three other words, *that* . . . is the stuff of poetry. What science and poetry share, even though they parted company, it seems centuries ago, is an honesty to the singular, determinate object. We tend to think science is after universals—the infinitely paraphraseable—to use again Guenther Stent's idea.[21] But science is not one thing, and maybe chemistry is different—we build shape, motion, and react on specific, variably persistent groups of atoms. Trends matter, general theories matter less. And individual molecules, examined up close, matter most of all.

Craving understanding, we circle around the object of our affections. In love with the particularity, the "thingness," of this powder, just this shade of turquoise, we study it. Here is what William Blake said:

> He who would do good to another, must do it in Minute Particulars:
> General Good is the plea of the scoundrel hypocrite & flatterer:
>
> For Art & Science cannot exist but in minutely organized Particulars.[22]

And the theme is voiced masterfully by A. R. Ammons, the American poet for whom art and science were not separated, in a section of his "Hymn":

> And I know if I find you I will have to stay with the earth
> inspecting with thin tools and ground eyes
> trusting the microvilli sporangia and simplest
> coelenterates
> and praying for a nerve cell
> with all the soul of my chemical reactions
> and going right on down where the eye sees only traces
>
> You are everywhere partial and entire
> You are on the inside of everything and on the outside
>
> I walk down the path down the hill where the sweetgum
> has begun to ooze spring sap at the cut

and I see how the bark cracks and winds like no other bark
chasmal to my ant-soul running up and down
and if I find you I must go out deep into your
        far resolutions
and if I find you I must stay here with the separate leaves.[23]

Fifteen million of the twenty million compounds known are white crystalline solids. I give you four vials, all white powders: one is sugar, another salt, the third penicillin, the fourth tetrodotoxin, the poison of the fugu or pufferfish. Will you play Russian roulette with these? Your body knows the difference. The difference, and its definition by the fallible powers of our mind and hands, is as beautiful as it is essential. The description of difference is one task the scientist does as well as it can be done.

Does the ethical bent inherent in the precision of language sought by scientists and poets make scientists and poets better human beings? No, no more than it improves those who professionally lead the considered life. The ethical impulse is strong, inherently human. It can be suppressed, most alarmingly by crowds and power, to use Canetti's phrase. And, remarkably enough, it can be suppressed by the flush of first creation. I'm thinking of the susceptibility to this of the saints: Sakharov and Bethe in science, Lowell and Sexton in poetry. Ethical thinking can be awakened; it needs to be reawakened, by consideration of whether a molecule can harm, of advances in reproductive technology, and . . . of just what one can invent in a historical play or whether a poem hurts a lover. Even a soap opera can teach ethics. We should be grateful for these little (or big) prods to ponder ethical choice.

The First (Fruitful) Intersection of Science and Ethics

I want to make a final point that returns to our cultural roots. The tree in the Garden of Eden in our primeval religious narrative was the Tree of Knowledge of Good and Evil. I take the *etz hadaat tov vera* as . . . "the Tree of Ethics" (a word not in ancient Hebrew), and the first link between science, narrative, and ethics. Let me bypass the question of why a just God would put ethics out of reach. He did. Or he didn't: if we knew good and evil just like that, the way we breathe, we wouldn't need ethics. Continuing in my disrespectful/respectful *midrash* (similar to that of

Zygmunt Bauman, Jean-Pierre Wils, and more recently Leon Kass),[24] is not Adam and Eve's transgression implicit in the tale, serpent or no serpent? Without it there would be no narrative, no story of humankind. We'd still be between the four rivers, right?

Even before eating of the fruit of the tree—and the rabbis discuss whether it was wheat, grapes, or fig, with no apple in sight—Eve makes a decision: "When the woman saw that the tree was good for eating and a delight for the eyes, and that the tree was desirable as a source of wisdom, she took of its fruit and ate."[25]

The science in my mildly sacrilegious *midrash* is manifold. It sparkles in the knowledge that the tree conveys. Of what? Of oppositions and polarities. Of choices, of course—of matter particulate and continuous, of opposites attracting each other or repelling, of analysis and synthesis. Of what is to be hidden and what is to be revealed, of the same and not the same (those four vials).

The narrative in Genesis 3:6 is so skillful that we rush to the denouement. But the sentence is worth reading again, and in the original Hebrew, so that there is no doubt as to the reasons given for Eve's actions. The verse speaks, directly, of experiment. For this is what Eve hazarded, isn't it? She saw, and thought, and acted. She acted on beauty, for wisdom. Kant would approve. Eve did what had to be done, not to end but to begin a story. In which curious human beings have the choice between good and evil.

Notes

Author's acknowledgments: Friends helped me a lot in reflecting on the topic of this essay—among them Sylvie Coyaud, Margery Arent Safir, and Emily Grosholz. I am grateful to Jennifer Cleland for her assistance with some research, and to M. M. Balakrishnarajan, Pradeep Gutta, and Beate Flemmig for some drawings.

1. Roald Hoffmann, "Fraudulent Molecules," *American Scientist* 85, no. 4 (1997), 314–17.
2. Daniel N. Wilson and Knud H. Nierhaus, "The Ribosome Through the Looking-Glass," *Angewandte Chemie, International Edition* 42, no. 30 (2003): 3464–86.
3. Peter C. Marzio, *Rube Goldberg: His Life and Work* (New York: Harper & Row, 1973).
4. Jerome S. Bruner, *Making Stories* (New York: Farrar, Straus, and Giroux, 2002), 90; and Roald Hoffmann, "Marginalia: Narrative," *American Scientist* 88, no. 4 (2000): 310–13. See also Hoffmann, "Why Buy That Theory?" *American Scientist* 91 (2002): 9–11.
5. L. B. Bjostad, C.E. Linn, J. W. Du, and W. L. Roelofs, "Identification of New Sex Pheromone Components in *Trichoplusia ni*, Predicted from Biosynthetic Precursors," *Journal of Chemical Ecology*

10, no. 9 (1984): 1309–23; and references therein; Julie L. Todd and Thomas C. Baker, "The Cutting Edge of Insect Olfaction," *American Entomologist* 43, no. 3 (1997): 174–82.

6. John C. Polanyi, "Science, Scientists and Society," *Queen's Quarterly* 107, no. 1 (2000): 31–36.
7. Walter Benjamin, "The Storyteller: Reflections on the Works of Nikolai Leskov," in *Illuminations*, ed. Hannah Arendt, trans. Harry Zohn, (New York: Harcourt, Brace & World, 1968), 90.
8. Hayden V. White, *The Content of the Form* (Baltimore: Johns Hopkins University Press, 1987), 23.
9. Emily Grosholz, personal communication by email, February 11, 2003.
10. Jacob Bronowski, *Science and Human Values*, rev. ed. (New York: Harper & Row, 1965), xiii.
11. Robert G. Bergman, "Irreproducibility in the Scientific Literature: How Often Do Scientists Tell the Truth and Nothing But the Truth?" *Perspectives On the Professions (CSEP)* 8, no. 2 (1989): 2–4.
12. Leonard Cassuto, "Big Trouble in the World of 'Big Physics,'" *Salon*, September 16, 2003, http://www.salon.com/2002/09/16/physics/.
13. T. S. Eliot, *Murder in the Cathedral* (New York: Harcourt, Brace & World, 1963), pt. 1, 44.
14. Kazuyuki Tatsumi et al., "Double Carbonylation of Actinide Bis(cyclopentadienyl) Complexes: Experimental and Theoretical Aspects," *Journal of the American Chemical Society* 108, no. 15 (1986): 4467–76.
15. Thomas S. Kuhn, *The Essential Tension* (Chicago: University of Chicago Press, 1977).
16. Anthony R. Pitochelli and M. Frederick Hawthorne, "The Isolation of the Icosahedral $B_{12}H_{12}^{-2}$ Ion," *Journal of the American Chemical Society* 82, no. 12 (1960): 3228–29.
17. Emily Grosholz, "Theomorphic Expression in Leibniz's *Discourse on Metaphysics*," *Studia Leibnitiana*, Thematic Issue: Towards the Discourse on Methaphysics, G. H. R. Parkinson, ed., 33, 1 (2001), 4–18; also personal communication by email, April 17, 2003.
18. Oliver Sacks, personal communication.
19. Roald Hoffmann, "Raisins for Being," in *Soliton: Poems* (Kirksville, MO: Truman State University Press, 2002), 28.
20. Soheila Chitsaz, Bernhard Neumüller, and Kurt Dehnicke, "Synthese und Kristallstruktur des gemischt-valenten Komplexes $[Sn_2I_3(NPPh_3)_3]$," *Zeitschrift für Anorganische und Allgemeine Chemie* 626, no. 4 (2000): 813–15. English translation: "Unusual Geometries and Questions of Oxidation State in Potential Sn(III) Chemistry," trans. Pradeep Gutta and Roald Hoffmann, *Inorganic Chemistry* 42, no. 25 (2003): 8161–70.
21. Guenther S. Stent, "Prematurity and Uniqueness in Scientific Discovery," *Scientific American* 227, no. 6 (1972): 84–93.
22. William Blake, *Jerusalem: The Emanation of the Giant Albion*, ed. Morton D. Paley (Princeton, NJ: William Blake Trust/Princeton University Press, 1991), 219.
23. A. R. Ammons, "Hymn," in *The Selected Poem: Expanded Edition*, ed. Andrew J. Porter, Henry L. Terrie Jr., and Edward J. Gordon (New York: Norton, 1986), 9.
24. Zygmunt Bauman, "What Prospects of Morality in Times of Uncertainty?" *Theory, Culture & Society* 15, no. 1 (1998): 11–22; and Jean-Pierre Wils, "Pleasure and Punishment: The Temptation of Knowledge," *Future* (2003): 74–80. See also Leon Kass, *The Beginning of Wisdom* (New York: Free Press, 2003).
25. Genesis 3:6. English translation in *The Torah* (Philadelphia: Jewish Publication Society, 1962).

Mieke Bal

The Ethics of Storytelling

WHO DOESN'T KNOW THE STORY OF JACOB'S SON JOSEPH, SEduced then betrayed by the wife of his boss, Potiphar? This story raises the issue of the bond between storytelling, ethics, and academic disciplines, in this case literary and theological analysis. If anything demonstrates the ethical dimension of cultural life, this story does because it embodies the kind of stories we tell about one another: gossip, slander, jokes, myths. All are representations that, by means of perpetuation and reiteration, establish or reconfirm ideas we have about other people. Some of these stories become canonical, and it is that canonical status that makes it all the more urgent to look at what they actually do.

We have been couching these concerns in terms of "identity" for some time now. The magic word *identity*, pointing to a feminist, queer, and/or multicultural perspective, and in itself connoting a progressive, politically inflected perspective, relates the need to subvert the exclusivist and oppressive power of canonical stories to the recognition that it is wielded by a small group of people with an identical identity. The word *identity* indicates that it is precisely the sociopolitical dominance of certain *kinds* of people over all "others" that the formation and maintenance of a canon facilitates.[1] But such a critique and revision of the canon goes hand in hand with a critique and revision of the ways we read the canon.

This other side of the double venture is indispensable if we are to understand the power of canons.[2]

There is more, however, to this larger project of opening up: it specifies a trajectory on which this cultural process of powerwielding functions in the story I have cited. Two cultural domains usually perceived as distinct, if not opposed—one worldly, the other spiritual—are related to each other. What we read for fun, enrichment, learning, or self-transformation contributes to the formation of who we are in terms of that specific cultural-political domain called *religion*.

There is a close link between artistic and religious canonicity, and the indisputable literary canonicity of an indisputably religious text like, say, Milton's *Paradise Lost*, proves it. This link serves an interest I wish to raise for discussion here before proceeding further. It is even more obvious in visual culture: the interest in *representation* as a sociopolitical tool. But, whereas I would emphasize that literature is close to visual art in its cultural "work," despite the fact that its medium is more acceptable to religions with a taboo on visual representation, I will limit myself for now to written texts.[3]

My analysis claims a *social agency* for literature. The need to rethink what this means has become urgent because of the social phenomenon of worldwide migration, which has both made the near-exclusive dominance of Christian and Judaic traditions in the Western world less obvious, and imposed participation in the social organization of—most visibly—Islamic traditions. The reality of migration has, more generally and vaguely but also more incisively, made us aware that religious identity is in fact not self-evidently Christian, or anything in particular at all.

Literature promotes exclusion and mainstreaming by means of its own canon and canonization. The latter are by no means identical, either in content or process, to the canons and canonization of established religions, and may therefore appear relatively "innocent" of imposing religious identity. However, precisely because religious canons are established once and for all, the more fluctuating literary canons may have even greater power to shape society.

Representation plays a key—but not exclusive—role in this power. As I will argue, "promotion" of and through literature is connected to representation in the sense of being a form of *incarnation*, or embodiment. I use the word *promotion* very much in film scholar Richard Dyer's sense—of promoting a particular way of considering the body as the site of, rather than defined by, spirit. As Dyer brilliantly observes in a critique of the tacit self-evidence of a predominantly "white" perception, this tradition insists on that curious notion of *incarnation* so central to Christianity. This tradition includes representing the body in suffering and sacrifice while, in terms of visual representation, countering the two other monotheistic traditions that resist visual representation of the body. Christianity has a stake in this that goes beyond just the wish to be liberal about art. Its primary stake, Dyer proposes, is the attempt to propagate a specific sense of embodiment of spirit. This is a specifically "white" endeavor to the extent that, if it is to be self-evidently predominant, the white body must be propagated but not foregrounded. Hence, these representations are about something that, in Dyer's words, is "in but not of the body."[4]

Ethical Nonindifference and Artistic Merit

By way of a revisionist exercise, I will now revisit an earlier analysis I published in my book *Reading "Rembrandt."*[5] This analysis proceeds from the perspective of the double knot—of identity and canon on the one hand, and religion and literature on the other. There I contrasted the biblical, canonical story of Joseph and Potiphar's wife with the equally canonical story in the Qur'an, of the "same" *fabula*. The issue was the representation of women, the term of comparison Thomas Mann's novel *Joseph and His Brothers*,[6] and the goal to demonstrate that even such oppressive and rigid mythical cases of women-bashing as Genesis 39 are not unified and need not be perpetuated.

The first issue I wish to recall here is the relation between *ethical* and *artistic* merit. When studying the conjunction of Rembrandt's representation of the story with the later literary version, I was struck by the fact that Mann, in his preface to *Joseph and His Brothers*, explicitly attempts to overcome the mythical reification of misogyny that he found in the traditions devoted to the episode. He is quite proud of "the humane vindication that I had undertaken in it, the humanisation of the figure of Potiphar's wife, the mournful story of her passionate love of the Canaanite major-domo and her *pro forma* husband."[7] On the strength of this uplifting political undertaking, Mann claims literary success for the episode, which he calls "unquestionably the artistic zenith of the work."[8] Importantly, then, he puts forward a protofeminist, *humanist* ideology as an argument for aesthetic accomplishment. What he claims, for his retelling of a story from the religious canon, is literary canonicity on ethical, political grounds.

Initially I was "reluctant to endorse such a conflation of aesthetics and ethics."[9] The reason for my reservation was the *ethical indifference* commonly shown in the face of artistic excellence. Such indifference has been displayed, for example, in the enthusiastic reception of wildly misogynist novels by Norman Mailer and Arthur Miller, or in the infamous case of Céline's brilliant anti-Semitic novels. My initial suspicion of ethical indifference was not wrong, and I still hold it. But I now wish to refine that view. I will not retrospectively endorse the *conflation*. Instead, I will revise, then promote, Mann's claim—but otherwise.

Usually this problem of the severed relation between ethics and aesthetics is resolved, albeit uneasily, by negating it. My solution in *Reading "Rembrandt,"* which focused only on the ethical element, now strikes me as escapist. Now at this time I would like to take the opposite position, and examine how ethical *nonindifference,* far from being indifferent to aesthetics, in fact informs this novel's artistic merit.

In the sentence from Mann's preface cited above, the vindication is already broached on two levels: that of the story and that of the fabula. I noted that "[b]y writing about him, Mann includes Potiphar's presence in the story . . . As a husband, Potiphar is only a signifier, an empty form, and it is the absence of the signified that triggers the story and makes it 'mournful.'"[10] *Then* I was sensitive to the understanding that Mann displayed for the wife's desire; that desire is the motivating force behind the fable. But the description of the character Potiphar, however brief, gives that desire a reason, a motivation, an excuse. "Mournful" gratified me as an acknowledgment of her plight at the story level.

Now I am sensitive to the underlying endorsement of monogamy, to the suggestion, "With such a husband, what can you expect?" As I have argued elsewhere, monogamy is not indifferent to either religious or, I now add, literary identity. Biblical metaphors of "whoring" for religious unfaithfulness prove it. More disturbingly, Mann's statement attributing the woman's desire to her husband's inadequacy deprives that desire of its "desirousness." An implicit moral consideration—such as excusing her adulterous desire because of the husband's deficiency—overrules the simple but profoundly subversive fact that the desire is both inalienably hers and at the same time weakens her identity—makes her, as the saying goes, weak-kneed. In other words, the figure of Mut-em-enet in Mann's novel, as well as the woman in the Qur'an version on which it draws, insists, more than Mann takes credit for in his preface, on a desire that is hers but that deprives her of her power to act socially and morally according to the rules. This may sound like another lame excuse, even lamer than her husband's inadequacy as a husband, but it is hers all the same. It is this site of the tangled relationship between elements that are usually distinguished—such as body and spirit, private and public—that I wish to foreground.

The first term I wish to examine is *body*, in the culturally specific sense of Christianity. Importantly, Mut's being weakened by desire is her state according to a conception of body that doesn't play the game with spirit that is promoted by Christianity in its narrative representations of embodiment of spirit. Think, for example, of the bodily iconography in Christian art, so thoroughly in opposition to Islamic and Judaic rules against such representations. Suffering and/as sacrifice are glorified through narratives like the *Pietà*,[11] which scene depicts nativity and death in conflation in order to convey the sense in which Christianity sees the bond between body and spirit. Mary is body in the sense of a vessel for spirit; Christ is body in which spirit is incarnated, placed; together they promote body as the site of embodiment. The antiquated Aristotelian gender division involved in the *Pietà* imagery is further enhanced—but again, *not* foregrounded, thus remaining unmarked—by the youthfulness of Mary in many such images. This suggests a heterosexual love relationship as the middle term between nativity and death. Representation, of the kind the other two monotheistic religions reject, is vitally important as a tool for promoting a sense of body that is specific, politically suited to enhancing domination, and paradoxically, at the same time, invisible because *unmarked*. Dyer argues that the representational *ideologeme* (my term, based on Jameson[12]) of incarnation invests unmarked white bodies with self-evident power through the conception of spirit that it implies, in combination with the political use to which that spirit is put, as "entrepreneurial" and imperialist.[13] This conjunction, between a Christian conception of body and imperialist politics, makes a critical reconsideration of both literary and religious canons, such as I am engaged in here, urgent, relevant, and timely.

My specific case study concerns desire. Why? Desire, I would think, is the systematic opposite of embodiment. It is the happening between *act* and *state* that is, in Dyer's terms, *of* the body, not localized *in* it. In view of this paradox I am interested in a scene where the *bodiliness* of desire and the impossibility of separating spirit and body become painfully clear, in the most literal sense of the adverb. It is the scene involving the little knives, which is absent from the biblical story. Extensively represented in Mann's novel, it is also found in that rival canon, the Qur'an. Mann's novel, then, does more than just vindicate the woman as if Mann were her ideological *cavalier servant*, turning her into a near-virgin by asserting her husband's inadequacy. In one sweep—a complex

but effective alternation of focalizing positions distributed between the woman Mut-em-enet, the Joseph figure named Osarsiph, and the other women in town—the novel accomplishes its literary greatness, its contribution to religious identity, and its opening up of the religious canon. Let me explain this assessment.

It is based on what happens following the woman's failed attempt to seduce Joseph—the scene, in other words, that vindicates her desire. To appreciate what Mann accomplishes here, it is important to acknowledge how this scene differs from the one in Genesis. There, after the fateful ultimatum that will take Joseph to prison, the nameless woman calls on "the men in the house" to slander him, using ethnic slurs to enlist the men's solidarity: "And . . . she called to the men of her household and said to them: 'See, he has brought among us a Hebrew to insult us; he came to me to lie with me and I cried out with a loud voice"' (Genesis 39:14).

In my earlier analysis I paid no attention to the woman's choice of arguments. It now strikes me that—to put it harshly for the sake of clarity—she appeals to racial sentiment and class subordination to denigrate her husband's household management in the eyes of his subordinates by fanning their envy of the higher-ranking Joseph.

The argument will be repeated to the husband who, without discussion, takes Joseph into custody. No wonder such a story has contributed to a further hardening of both misogynous mythology and—not to be forgotten—the antagonism between the intended Hebrew readers of the story and their "others," to whom this woman ethnically belongs. Note that it is the husband whom she accuses of disrespect for her and for the other men in the house. The Hebrew man is just an instrument, as are the other servants. The woman uses race and class to get round to sex. Conversely, the story uses sex to condemn femininity, while promoting—"self-evidently," without foregrounding—race and class putdowns, as well as the straitjacket of male bonding and solidarity in a hierarchically organized society. The excuse that a humanist, feminist perspective could offer is that the woman's strategy utilizes the only instruments available to a woman confined to a house in a predominantly male world.

But Mann's ethical nonindifference is artistically motivated, just as the artistic quality of his work is ethically nonindifferent. Although I would need to write far more to substantiate this claim, each of the three stories, if framed within its own aesthetics, has literary brilliance. And in each case, the literary quality involves ethical issues. Because the means of, especially, narrative literature touches upon the intricate connections between private and public aspects of subjectivity, attempts to pass off objectionable ethics under the cover of literary brilliance are doomed. So are attempts to separate the two—that is, to condemn a work on ethical grounds while continuing to read it canonically on aesthetic premises—through ethical indifference.

I will touch on aesthetic concerns in Mann's text only. These concerns are explicit in the author's preface and also known to the readers of the novel, who share the kind of canonical form of literature within which the novel was conceived, written, and read. The issue at hand in this novel is *realism*. But to write a novel—hundreds of pages, realistic detail, psychology, and all that that entails—Mann's vindication needs a richer source. As Mann complained, the biblically meager "what" lacked the "how" and "why": life's circumstantial detail was missing. So, at first sight, this is a literary issue, and what Mann gives expression to is the novel's realistic needs. Circumstantial detail is a notable feature of realistic narrative, and an important element of aesthetic judgments of such narratives, which may even decide their inclusion or exclusion from the canon.[14]

Realism in literature and art requires detail. Remember, though, that realism as a *belief* in the reality of representation is also the ground for the iconophobia of orthodox Judaism and Islam. I submit that these two aspects knot together religious and literary concerns. What leads to *idolatry* in religion leads to *sympathy* in the novel, and both are equally dangerous lures to transgression. The difference is the conception of the body—as the *site* of embodied spirit rather than as *defined* by spirit. This bond between religious and literary taboos on the one hand, and between conceptions of body and spirit on the other, becomes clearer when we realize that there is also a legal point in Mann's complaint. Circumstantial evidence, although legally flimsy and problematic, is often the only evidence there is with which to indict or acquit. Hence, the artistic consideration—how to gain a place in the literary canon by doing an aesthetically good job with details—*inevitably* entails considerations that spill over into the social domain, where politics and ethics meet in the face of the law. And not to be forgotten is the sense in which the Greek word *kanon* means "law."

Mann's search for circumstantial detail in a story where guilt is at stake is, therefore, in and of itself an act of merging aesthetics and ethics; an act, that is, *against the ethical indifference of aesthetics*. This is why his statement in the preface deserves endorsement, say, as a philosophical position. Against a Kantian disinterestedness, Mann, on the eve of unimaginable crimes against humanity, postulates an aesthetics of ethical nonindifference.

The Cutting Edge of Literary Identity

I will now argue two points that result from the need for ethical nonindifference as an element of aesthetics. First, *literariness* is a tool for identity formation. Second, religious canonicity is not premised on that formation but rather allows and even facilitates it. I will argue these points by following Mann in his search for detail.

Mann finds his circumstantial evidence in the rival text—the later, so to speak, derivative version: the episode that appears in the Qur'an

12:23–35, where the follow-up of the seduction attempt is different and more extended. Yet the Qur'an is not a novel, any more than Genesis is. While adding "detail," some of the circumstantial features in Genesis are also eliminated, which is why I find the terms "gap" and "gap-filling" unhelpful. They fail to account for repression and elimination, and end up accounting only for supplementation. There is no equivalent to Genesis 39:14, no interaction with the other men in the house, no racial slurs, no appeal to class subordination to solicit sympathy.

This omission, especially of ethnic denigration, is easily explained in view of the respective ideological and religious identities of the intended readers, and of the context within which these texts function. I am certainly not interested in further exacerbating the divide between Judaism and Islam, the respective curators of these canonical legacies.[15] I only mention the omission so as to footnote Mann's seemingly *quantitative* complaint, for the appeal to support and solidarity, which in Genesis carries the "argumentation" that we have already seen, features in the Qur'an as well. It is only the people called upon who are different. The social setting shifts. Mann's choice of the Qur'an episode to aesthetically supplement the meager Hebrew text is therefore not at all a supplementation or gap filling, but instead stems from ethical concern; it is an act of ethical nonindifference.

This choice highlights an important difference between the two religious texts: in the Qur'an, the woman is not the only woman in the story. There, the woman becomes subject to slander from her women friends "in the City" (12:30). The scene is twice displaced—from men to women and from house to city. Significantly, the woman acquires a public status that enables her to act. And although, according to the fabula, she pulls the same trick on Joseph as in Genesis, and hence lends herself to the same misogynous mythologizing, the story is given an instructively different twist that prevents such stabilization. The women in the city assume that the woman is going crazy from love (12:30). She then acts, not to disqualify or harm Joseph, or to deny her desire, but to gain literal *sympathy*: co-suffering from desire. Here is the passage from the Qur'an (12:31):

> When she heard
> Of their malicious talk
> She sent for them

And prepared a banquet
For them: she gave
Each of them a knife:
And she said (to Joseph),
"Come out before them."
When they saw him,
They did extol him,
And (in their amazement)
Cut their hands: they said,
"God preserve us! No mortal
Is this! This is none other
Than a noble angel!"

She then proceeds to confess to her friends that she did in fact seduce and trick him, and that he did "firmly save himself guiltless" (12:32). Potiphar and "the men" in fact imprison Joseph for his own good, "for a time" (12:35).

When I first studied this case, I was interested in Mann's use of the Qur'an to vindicate the woman. Now I wonder what his choice of that text offers more specifically in terms of the bond between storytelling, ethics, and academic disciplines. As far as literary status is concerned, first and most obviously, there is more to the historical position of Mann's text than I realized, especially in terms of ethnicity, one extremely important element of identity. Moreover, this historical consideration—the novel's pre-Holocaust position—needs to be juxtaposed with my own position, speaking from a post-Holocaust but also "global" period of migration and cultural mixing.

Second, as we see below, Mann's novel, by means of a (literary, realistic) detail—the oranges—adds more *pain*. Remember what I said earlier about suffering and sacrifice. Literary identity as bound up with representation is here elaborated in terms of, yet in critical engagement with, Christian ideology. Here pain is *not* an overcoming of the body by a spirit only temporarily housed there, but an endorsement of an undivided body of desire. In other words, here physical pain is of such a nature that, through its very metaphorical affirmation of the woman's love pain, it transforms desire from moral-political to bodily, in an acceptance that spirit and body are one.

I did realize this pain was important when I wrote:

> At dessert, oranges and very sharp little knives to peel them are distributed. And when the ladies are busy peeling their oranges, Osarsiph/Joseph comes in to serve wine. Sheer fright at the *sight* of such beauty makes all the women cut their fingers, some of them to the bone. "My loves, what ever has happened to you all? What are you doing? Your blood is flowing" is the reaction of the plotting, lovesick virgin. Her intentionally insincere exclamation receives sincerity from the contiguous comment in which the speaker fails to distinguish between himself and this woman. For the narrator continues: "It was a fearful sight."[16]

Now, however, both this sharper pain—caused by the acidity of the oranges—and its bloody visibility appear more striking in terms of theorizing identity through identification, as one form of "literary identity." This is played out aesthetically on the level of the detail of representation, and ethically on the level of shifting the interpretation of the body to confront it with Christian conceptions.

Two points emerge here. First, literariness is the tool for identity formation. Second, religious canonicity is not premised on identity formation but allows, even facilitates it. Concerning the first point, Mann adequately walks the tightrope between ethical and aesthetic work by mediating on the level of imagination. He does this by practicing a literary genre in which evaluation of a character's actions is much more common than it was in ancient texts, but which aesthetically calls for such mitigating and complicating tropes as irony and such narrative devices as alternating, double, and ambiguous focalization. Having created an imaginary realm that readers and characters can share, the appeal to other subjects is here also an appeal to other readers.

The same happens on the level of the fable. The women friends feel the pain of their friend's desire, when they feel it themselves at the moment *they* are hurt. This common pain produces community—literal, bodily community. Common ground is produced through heteropathic identification, to use Kaja Silverman's term (a form of identification where the subject, rather than absorbing the other's alterity to make her the same, extols it, idealizes it, and goes out of herself to share it).[17] This is the meaning of the plural as well as of "City" in "ladies of the City."

Although the banquet takes place in the confined space of Mut's house, the public life of the city enters that space, opening its walls and liberating her from confinement, and thus facilitating contact.

The Qur'an text already did something similar; Mann found there his circumstantial evidence, to which the oranges, as a modern "effect of the real," were easy to add.[18] But the point that I was unable to see when I first studied this case is the importance of the conjunction of *pain* with *women* and *City* in this version. The recognition of women's public life is as important as their frail, threatened, but potentially salvaged, solidarity on the basis of something perceived so deceptively as private as desire. It is in this semantic space that the identity produced—on the imaginary level of literature—can be mobilized for, perhaps, a "religious" identity. Not religious as in Judaic, Christian, or Islamic official, regulated religion, but in another sense.

This identity—to put it a bit flippantly, of women collectively, hence publicly, entitled to their desires, even if this does not accord with public morality—has no canonical status. But the texts that offer it as a possibility do: Mann's novel in a secularized world has literary canonicity; the Qur'an in its religious function has canonical status. The two cases belong to their respective mainstreams—although it seems important to note that these two mainstreams are quite separate—and it would seem a fair guess to state that their canonical status is *not* due to their "vindication of the rights of women to desire." In other words, if you agree that this right to desire is ethically "good," it follows that neither the literary nor the religious canonicity can be confined to ethical "goodness" or "badness." Nor is the biblical story's canonical status imputable to the disturbing sentiments its more clearly wicked woman character solicits to promote her case.

My provisional conclusion, therefore, is the following. Literary merit, as modern Western culture has construed it and on which it has based its canonization process, is not concerned with good or evil in themselves, for these are culturally and historically specific, but with *ethical nonindifference*. Hence, ethical nonindifference contributes to potential literary canonicity. But as my case study clearly demonstrates, the canonization as either literary masterpiece or religiously institutionalized scripture is not directly based on the particular ethical merit of the positions represented.

The second point I seem to have been blind to *then* and which I find important *now* (although I still feel ill-equipped to address it) concerns the element "religion." None of the passages featuring Potiphar's wife are themselves explicitly religious, although I have argued that the scene with the little knives does revisit Christian ideology critically. One would instead be inclined to say, at least for the two religious canons, that these passages are ancient myths that have ended up being canonized for some reason that may have nothing to do with the respective laws of Judaism and Islam. This very plausible lay view of things has made it easy for me to discuss biblical texts in nontheological terms. In the historical era of global migration in which we are living, religion cannot be bracketed as easily as seemed possible then.

If religion is what binds cultural subjects, it could well be one of those domains that mediates between the terms of the infelicitous but persistent binary oppositions of private/public, individual/collective, body/spirit, and the like. With that understanding, if the canonical status of at least the Qur'an version may offer something that is not "merely" literary, what can the subversive literary identity of our woman figure(s) mean in "religious" terms?

Double Binding: Literary Identity and Its Religious Effects

Indeed, what is it that would make the triple case of Potiphar's wife's mishaps (her tricks, her powerlessness, her desire) "religious," in the sense of offering an imaginary realm where such oppositions can be suspended, a realm that would be forbidden if the imagination were visually represented?

First, the pain caused by the cutting knives (or was it the shared desire?) and exacerbated by the acidic oranges (or was it the colorful detail?) is a metaphor: "Pain" translates the way the irreducible difference and the subsequent loneliness of an individual, especially one not quite firmly positioned in mainstream society, *cuts*; it severs and hurts.

But, second, the blood thus shed also stands for the hot blood of desire that is, as such, the movement out of the autistic self into a risky, dangerous but fulfilling, and, indeed, indispensable communication with another. Third, then, the pain, or *pathos*, is a *sym-pathos*, a pain that as pain binds, by means of heteropathic identification, the lonely woman, sick from unrequited love, with her friends who, a moment before, had dropped her like a hot potato into the well of social isolation. For me the blood also symbolizes a multiple "marriage," the emotional defloration of the women enslaved by imposed monogamy and potentially accepting the risks of their own desires in a social sphere where they are not alone with the one man who owns them. Through the knives, their woman-friend "penetrates" their bodies, releasing the blood of the ruptured hymen.

Theirs is a binding that does not cancel but only temporarily suspends the self-evidence of monogamy—and perhaps even of its metaphorical counterpart, monotheism. Now, is that religion? I don't know, but the place where I see that possibility opened up is part of a religious canon that as a whole is indifferent to such issues, and such plights. *Two forms of binding can coexist.* This seems important in a culture—ours, today—where respect for difference is often confused with indifference.

Thus, the age-old story of this nameless female trickster was abducted from a double religious canon—the Bible and the Qu'ran—to become an aesthetic object, only to serve as another form of binding that does not require canonicity. Mann produced a female figure, at history's most fraught and violent time, as his contribution to life-with-desire because it is of the body. To perform that feat, he proceeded to undercut the close bond between monotheism, monogamy, and masculinity by depicting, for our imagination, the "adulterous" woman as a figure who produces—in her rather nasty, painful agency, limited but also enabled by institution and framing—the condition that permits us to think canons, religious or not, differently.

Notes

1. In this sense the present theme forms part of a larger project well underway since the late 1970s. In 1981, an institution itself as canonical as the annual English Institute held in the United States, August 31–September 3, 1979, produced a famous collection of essays called *English Literature, Opening Up the Canon*, ed. Lesley A. Fiedler and Houston A. Baker (Baltimore: Johns Hopkins University Press, 1981).

2. This is not only a political necessity but also, and even primarily, an intellectual and academic one, for understanding is the "official," "canonical" task of the academic endeavor.

3. The reason I insist on this affiliation between the two arts is that the prohibition of making graven images has realism, not vision *per se*, as its target.

4. Richard Dyer, *White* (London: Routledge, 1997).

5. Mieke Bal, *Reading "Rembrandt": Beyond the Word-Image Opposition* (Cambridge: Cambridge University Press, 1991). An extended version of that analysis has meanwhile has been published as *Loving Yusuf: Conceptual Travels from Present to Past* (Chicago: University of Chicago Press, 2008).

6. Thomas Mann, *Joseph and His Brothers*, trans. H. T. Lowe-Porter (New York: Alfred A. Knopf; Harmondsworth: Penguin, 1948).

7. Ibid., xi.

8. Ibid.

9. Bal, *Reading "Rembrandt,"* 97.

10. Ibid., 96.

11. Elaine Scarry, *The Body in Pain* (Oxford: Oxford University Press, 1985), 216.

12. Fredric Jameson, *The Political Unconscious* (Ithaca, NY: Cornell University Press, 1981).

13. Dyer, *White*, 14–40.

14. For the aesthetics and ethics of detail, see Naomi Schor, *Reading in Detail* (New York: Methuen, 1987).

15. For an important argument in favor of this restraint, see Edward W. Said, "Invention, Memory, and Place," *Critical Inquiry* 26, no. 2 (2000): 175–92. For an analysis of similar symbols in opposed traditions, see also Carol B. Bardenstein, "Trees, Forests, and the Shaping of Palestinian and Israeli Collective Memory," in *Acts of Memory: Cultural Recall in the Present*, ed. Mieke Bal, Jonathan V. Crewe, and Leo Spitzer (Hanover, NH: Dartmouth College, University Press of New England, 1999), 146–68.

16. Mann, *Joseph and His Brothers*, 803, quoted in Bal, *Reading "Rembrandt,"* 97.

17. Kaja Silverman, *The Threshold of the Visible World* (New York: Routledge, 1996).

18. Roland Barthes, "L'effet du réel," *Communications* 11, no. 11 (1968): Recherches sémiologiques le vraisemblable, 84–89. English translation: "The Reality Effect," in *The Rustle of Language*, trans. Richard Howard (New York: Hill and Wang, 1986), 141–54.

Works Cited

Ammons, A. R. "Hymn." In *The Selected Poem: Expanded Edition*, edited by Andrew J. Porter, Henry L. Terrie Jr., and Edward J. Gordon, 9. New York: Norton, 1986.

Aristotle. *Historia animalium*. Translated by A. L. Peck. 2 vols. Loeb Classical Library. Cambridge, MA: Harvard University Press, 1970.

——. *On Poetics*. In *Great Books*. Translated by W. Rhys Roberts. vol. 8. Chicago: Encyclopædia Britannica, 1993.

Bal, Mieke. *Reading "Rembrandt": Beyond the Word-Image Opposition*. Cambridge: Cambridge University Press, 1991.

——. *Loving Yusuf: Conceptual Travels from Present to Past* (Chicago: University of Chicago Press, 2008).

Bardenstein, Carol B. "Trees, Forests, and the Shaping of Palestinian and Israeli Collective Memory." In *Acts of Memory: Cultural Recall in the Present*, edited by Mieke Bal, Jonathan Crewe, and Leo Spitzer, 146–68. Hanover, NH: Dartmouth College, University Press of New England, 1999.

Barrow, John D. *The Artful Universe: The Cosmic Source of Human Creativity*. Oxford: Oxford University Press, 1995.

——. *Pi in the Sky: Counting, Thinking, and Being*. Oxford: Clarendon, 1992.

Barthes, Roland. *Critical Essays*. Translated by Richard Howard. Evanston, IL: Northwestern University Press, 1972.

——. "L'effet du réel." *Communications* 11, no. 11 (1968): Recherches sémiologiques le vraisemblable, 84–89.

——. *Essais critiques*. Paris: Editions du Seuil, 1964.

——. *Mythologies*. Paris: Editions du Seuil, 1957.

——. *Mythologies*. Translated by Annette Lavers. New York: Hill and Wang, 1972.

——. "The Reality Effect." In *The Rustle of Language*, translated by Richard Howard, 141–54. New York: Hill and Wang, 1986.

Bauman, Zygmunt. "What Prospects of Morality in Times of Uncertainty?" *Theory, Culture & Society* 15, no. 1 (1998): 11–22.

Benjamin, Walter. "The Storyteller: Reflections on the Works of Nikolai Leskov." In *Illuminations: Essays and Reflections*, edited by Hannah Arendt, translated by Harry Zohn, 83–109. New York: Harcourt, Brace & World, 1968.

Bergman, Robert G. "Irreproducibility in the Scientific Literature: How Often Do Scientists Tell the Truth and Nothing But the Truth?" *Perspectives* 8, no. 2 (1989): 2–4. http://ethics.iit.edu/perspective/v8n2%20perspective.pdf.

Bjostad, L. B., C. E. Linn, J.-W. Du, and W. L. Roelofs. "Identification of New Sex Pheromone Components in *Trichoplusia ni*, Predicted from Biosynthetic Precursors." *Journal of Chemical Ecology* 10, no. 9 (1984): 1309–23.

Blake, William. *Jerusalem: The Emanation of the Giant Albion*. Edited by Morton D. Paley. Princeton, NJ: William Blake Trust/Princeton University Press, 1991.

Borges, Jorge Luis. "Avatars of the Tortoise." In *Labyrinths: Selected Stories and Other Writings*, translated and edited by Donald A. Yates and James E. Irby, 202–8. New York: New Directions, 1964.

——. "La Biblioteca de Babel." In *Obras completas 1923–1949*, 465–71. Buenos Aires: Emecé, 1939. Reprint 1989.

——. "Kafka y sus precursores." In *Otras inquisiciones*, 145–48. Buenos Aires: Emecé, 1951–1952. Reprint 1960.

——. "The Library of Babel." In *Labyrinths: Selected Stories and Other Writings*, translated by James E. Irby, 51–58. New York: New Directions, 1964.

——. "Pierre Menard, Author of the *Quixote*." In *Labyrinths: Selected Stories and Other Writings*, translated by James E. Irby, 36–44. New York: New Directions, 1964.

——. "Pierre Menard, autor del *Quijote*." In *Obras completas 1923–1949*, 444–50. Buenos Aires: Emecé, 1939. Reprint 1989.

Bronowski, Jacob. *Science and Human Values*. rev. ed. New York: Harper & Row, 1965.

Bruner, Jerome S. *Making Stories: Law, Literature, Life*. New York: Farrar, Straus, and Giroux, 2002.

Carroll, Lewis. "Jabberwocky." In *Through the Looking-Glass*. London: Macmillan, 1871.

Cassuto, Leonard. "Big Trouble in the World of 'Big Physics.'" *Salon*. September 16, 2003. http://www.salon.com/2002/09/16/physics/.

Cavell, Stanley. *Pursuits of Happiness: The Hollywood Comedy of Remarriage*. Cambridge, MA: Harvard University Press, 1981.

Chitsaz, Soheila, Bernhard Neumüller, and Kurt Dehnicke. "Synthese und Kristallstruktur des gemischt-valenten Komplexes $[Sn_2I_3(NPPh_3)_3]$." *Zeitschrift für Anorganische und Allgemeine Chemie* 626, no. 4 (2000): 813–815.

——. "Unusual Geometries and Questions of Oxidation State in Potential Sn(III) Chemistry." translated by Pradeep Gutta and Roald Hoffmann. *Inorganic Chemistry* 42, no. 25 (2003): 8161–70.

Cortázar, Julio. *Hopscotch*. Translated by Gregory Rabassa. New York: Pantheon, 1966.

——. *Rayuela*. Buenos Aires: Sudamericana, 1963.

Dante Alighieri. *Inferno: The Divine Comedy of Dane Alighieri*. Translated by Allen Mandelbaum. New York: Bantam, 1982.

Davies, Paul. *God and the New Physics*. London: Penguin, 1983. Reprint 1990.

Dear, Peter, ed. *The Literary Structure of Scientific Argument: Historical Studies*. Philadelphia: University of Pennsylvania Press, 1991.

Dyer, Richard. *White*. London: Routledge, 1997.

Eliot, T. S. *Murder in the Cathedral*. New York: Harcourt, Brace & World, 1963.

Faure, Paul. *Ulysse le Crétois : XIIIe siècle avant J.-C.* Paris: Fayard, 1980.

Fiedler, Leslie A., and Houston A. Baker, eds. *English Literature, Opening Up the Canon: Selected Papers from the English Institute, Conference held on August 31 to September 3, 1979*. Baltimore: Johns Hopkins University Press, 1981.

Gell-Mann, Murray. *The Quark and the Jaguar: Adventures in the Simple and the Complex*. New York: W. H. Freeman, 1994.

Gleick, James. *Chaos: Making a New Science.* New York: Viking Penguin, 1987.

Gross, Alan G. *Rhetoric of Science.* Cambridge, MA: Harvard University Press, 1990.

Gross, Alan G., and William M. Keith, eds. *Rhetorical Hermeneutics: Invention and Interpretation in the Age of Science.* Albany: State University of New York Press, 1997.

Guitton, Jean, Grichka Bogdanov, and Igor Bogdanov. *Dieu et la science: vers le métaréalisme.* Paris: Grasset, 1991.

Gutta, Pradeep, and Roald Hoffmann, trans. "Unusual Geometries and Questions of Oxidation State in Potential Sn(III) Chemistry," by Soheila Chitsaz, Bernhard Neumüller, and Kurt Dehnicke. *Inorganic Chemistry* 42, no. 25 (2003): 8161–70.

Heidegger, Martin. *Nietzsche.* 2 vols. Pfullingen: Günther Neske, 1961.

Heraclitus. *Fragments: Translations and Analysis,* by Dennis Sweet. Lanham, MD: University Press of America, 1995.

Hesse, Mary B. "Models, Metaphors and Truth." In *Knowledge and Language. Metaphor and Knowledge,* edited by Frank R. Ankersmit and Jan Johann Albinn Mooij, vol. 3, 49–66. Dordrecht: Kluwer Academic Publishers, 1993.

——. "Tropical Talk: The Myth of the Literal." *Aristotelian Society* 61, supp. (1987): 297–311.

Hoffmann, Roald. "Fraudulent Molecules." *American Scientist* 85, no. 4 (1997): 314–17.

——. "Marginalia: Narrative." *American Scientist* 88, no. 4 (2000): 310–13.

——. "Raisins for Being." In *Soliton.* Kirksville, MO: Truman State University Press, 2002.

——. "Why Buy That Theory?" *American Scientist* 91, no. 1 (2002): 9–11.

Jameson, Fredric. *The Political Unconscious: Narrative as a Socially Symbolic Act.* Ithaca, NY: Cornell University Press, 1981.

Jastrow, Robert. *The Enchanted Loom: Mind in the Universe.* New York: Simon and Schuster Trade, 1981.

——. *God and the Astronomers.* New York: W.W. Norton, 1992.

Jewish Publication Society of America. *The Torah: The Five Books of Moses.* Philadelphia: Jewish Publication Society, 1962.

Judson, Horace Freeland. *The Eighth Day of Creation: Makers of the Revolution in Biology.* London: Penguin, 1994.

Kass, Leon. *The Beginning of Wisdom: Reading Genesis.* New York: Free Press, 2003.

Keller, Evelyn Fox. *Making Sense of Life: Explaining Biological Development with Models, Metaphors, and Machines.* Cambridge, MA: Harvard University Press, 2002.

Kuhn, Thomas S. *The Essential Tension: Selected Studies in Scientific Tradition and Change.* Chicago: University of Chicago Press, 1977.

Lacan, Jacques. *Le Seminaire VIII: Le Transfert.* Paris: Seuil, 1991.

Lecercle, Jean-Jacques. "Lewis Carroll et le Talmud." *Théorie, Littérature, Enseignement* 10 (1992): 169–87.

Lucian, of Samosata. *Works.* Translated by A. M. Harmon. vol. 3 of 8, Loeb Classical Library. Cambridge, MA: Harvard University Press, 1995.

Mann, Thomas. *Joseph and His Brothers*. Translated by H. T. Lowe-Porter. New York: Alfred A. Knopf; Harmondsworth: Penguin, 1948.

Marzio, Peter C. *Rube Goldberg: His Life and Work*. New York: Harper & Row, 1973.

Merriam-Webster's Dictionary of English Usage. Springfield, MA: Merriam-Webster, n.d. http://www.merriam-webster.com/dictionary/neologism.

Nietzsche, Friedrich. *Also sprach Zarathustra*. Münich: Goldmann Verlag, s.d.

——. *The Birth of Tragedy and Other Writings*. Edited by Raymond Geuss and Ronald Speirs, translated by Ronald Speirs. Cambridge: Cambridge University Press, 1999.

——. *Dernières lettres*. Translated by Catherine Perret. Paris: Rivages, 1989.

——. *Human, All Too Human*. Translated by R. J. Hollingdale. Cambridge: Cambridge University Press, 1986.

——. *Introduction théorétique sur la Vérité et le Mensonge au sens extra-moral*. In *Le Livre du philosophe: études théoriques*, translated by Angèle. Kremer-Marietti, 117–33. Paris: Aubier-Flammarion, 1969. Reprint 1991.

——. *On Truth and Lying in an Extra-Moral Sense*. In *Friedrich Nietzsche on Rhetoric and Language*, edited and translated by Sander L. Gilman, Carole Blair, and David J. Parent, 246–57. New York: Oxford University Press, 1989.

——. *Thus Spoke Zarathustra: A Book for All and None*. Translated by Walter Kaufmann. New York: Penguin, 1978.

——. *Über Wahrheit und Lüge: ein Essay, Aphorismen und Briefe*. Edited by Steffen Dietzsche. Frankfurt: Insel, 2000.

——. *Unpublished Writings from the Period of Unfashionable Observations. Complete Works*. Translated by Richard T. Gray. Stanford, CA: Stanford University Press, 1995.

——. *Untimely Meditations*. Edited by Daniel Breazeale, translated by R. J. Hollingdale. Cambridge: Cambridge University Press, 1997.

——. *Unzeitgemäße Betrachtungen*. Munich: Goldmann, 1964.

——. *Werke; Kritische Gesamtausgabe*, Edited by Giorgio Colli and Mazzino Montinari. sec. 3, vol. 4. Berlin: Walter de Gruyter, 1978.

"Non-scientific readers of science." *The Economist*, May 9, 1998, 97–98. http://www.economist.com/node/127818.

Oxford English Dictionary. 2nd ed., 2 vols., s.v. Oxford: Oxford University Press, 1936.

Pitochelli, Anthony R., and M. Frederick Hawthorne. "The Isolation of the Icosahedral $B_{12}H_{12}^{-2}$ Ion." *Journal of the American Chemical Society* 82, no. 12 (1960): 3228–29.

Plato. *The Collected Dialogues of Plato, including the letters*. Edited by Edith Hamilton and Huntington Cairns. Princeton, NJ: Princeton University Press, 1996.

Polanyi, John C. "Science, Scientists and Society." *Queen's Quarterly* 107, no. 1 (2000): 31–36.

Rabaté, Jean-Michel. *The Ghosts of Modernity*. Gainesville: University Press of Florida, 1996.

——. *James Joyce and the Politics of Egoism*. Cambridge: Cambridge University Press, 2001.

——. *La Pénultième est morte: Spectographie de la modernité*. Seyssel: Champ Vallon, 1993.

Rio, Michel. "L'essentiel et l'accessoire." In *Rêve de logique: essais critiques*. Paris: Editions du Seuil, 1992.

———. *La Terre gaste: Roman*. Paris: Editions du Seuil, 2003.

Safir, Margery Arent, ed. "Introduction." In *Melancholies of Knowledge: Literature in the Age of Science*, 1–25. Albany: State University of New York Press, 1999.

Said, Edward W. "Invention, Memory, and Place." *Critical Inquiry* 26, n. 2 (2000): 175–92.

Scarry, Elaine. *The Body in Pain: The Making and Unmaking of the World*. Oxford: Oxford University Press, 1985.

Scheler, Max. *Ressentiment*. Edited by Lewis A. Coser, translated by William W. Holdheim,. New York: Free Press of Glencoe, 1961.

———. "Das Ressentiment im Aufbau der Moralen." In *Gesammelte Werke*. vol. 3 of 15, 33–148. Berne: Francke Verlag, 1955.

Schor, Naomi. *Reading in Detail: Aesthetics and the Feminine*. New York: Methuen, 1987.

Selzer, Jack, ed. *Understanding Scientific Prose*. Madison: University of Wisconsin Press, 1993.

Serres, Michel. *Feux et signaux de brume, Zola*. Paris: Grasset, 1975.

Shell, Marc. *Children of the Earth: Literature, Politics and Nationhood*. New York: Oxford University Press, 1993.

Silverman, Kaja. *The Threshold of the Visible World*. New York: Routledge, 1996.

Stent, Guenther S. "Prematurity and Uniqueness in Scientific Discovery." *Scientific American* 227, no. 6 (1972): 84–93.

Strosberg, Eliane. *Art and Science*. Paris: UNESCO, 1999.

Tatsumi, Kazuyuki, Akira Nakamura, Peter Hofmann, Roald Hoffmann, Kenneth G. Moloy, and Tobin J. Marks. "Double Carbonylation of Actinide Bis(cyclopentadienyl) Complexes: Experimental and Theoretical Aspects." *Journal of the American Chemical Society* 108, no. 15 (1986): 4467–76.

Taylor, Charles Alan. *Defining Science: A Rhetoric of Demarcation*. Madison: University of Wisconsin Press, 1996.

Thomas, Lewis. *Late Night Thoughts on Listening to Mahler's Ninth Symphony*. New York: Bantam Books, 1984.

Todd, Julie L., and Thomas C. Baker. "The Cutting Edge of Insect Olfaction." *American Entomologist* 43, no. 3 (1997): 174–82.

Trinh, Xuan Thuan. *La Mélodie secrète: Et l'Homme créa l'Univers*. Paris: Fayard, 1988.

U.S. Department of Energy. Genomic Science Program. *Notes on the Visions for Computational and Systems Biology Workshop for the Genomes to Life Program*. Washington, DC, September 6–7, 2001. http://doegenomestolife.org/pubs/CompbioVisions-4.pdf.

Vico, Giambattista. *La Science nouvelle*. Translated by Ariel Doubine, edited by Fausto Nicolin. Paris: Nagel, 1953.

White, Hayden V. *The Content of the Form: Narrative Discourse and Historical Representation*. Baltimore: Johns Hopkins University Press, 1987.

Wils, Jean-Pierre. "Pleasure and Punishment: The Temptation of Knowledge." *Future* (2003): 74–80.

Wilson, Daniel N., and Knud H. Nierhaus. "The Ribosome Through the Looking-Glass." *Angewandte Chemie, International Edition* 42, no. 30 (2003): 3464–86.

Wittgenstein, Ludwig. *Philosophical Investigations*. Translated by G. E. M. Anscombe. 3rd ed. New York: Macmillan, 1970.

——. *Philosophische Untersuchungen*. Frankfurt: Suhrkamp, 1967.

Zidi, Claude, Simon Michaël, Didier Kaminka, and James Cameron. *True Lies*. Directed by James Cameron (July 15, 1994). Los Angeles: Twentieth Century Fox, 1994. Based on Zidi, Claude, Simon Michaël, and Didier Kaminka. *La Totale!* Directed by Claude Zidi (December 18, 1991). Paris: AMLF, 1991.

Index

About the Contributors

MIEKE BAL is a cultural theorist and critic, based at the Amsterdam School for Cultural Analysis (ASCA), University of Amsterdam. Her areas of interest range from biblical and classical antiquity to seventeenth-century and contemporary art, modern literature, feminism, and migratory culture. Her many books include *A Mieke Bal Reader* (2006), *Travelling Concepts in the Humanities* (2002), and *Narratology* (1985), now in its third edition. She is also a video artist. Her internationally exhibited documentaries on migration include *Separations, State of Suspension, Becoming Vera,* and the installation *Nothing is Missing*. With Michelle Williams Gamaker, Bal made the feature film *A Long History of Madness*, a theoretical fiction about madness, and related exhibitions (2012). Occasionally, she acts as an independent curator, and her co-curated exhibition 2MOVE has traveled to four countries. Her current project *Madame B: Explorations in Emotional Capitalism*, also with Michelle Williams Gamaker, is exhibited worldwide. She is beginning a dual-screen installation about René Descartes and Queen Christina of Sweden.

ROALD HOFFMANN is the Frank H. T. Rhodes Professor Emeritus of Humane Letters at Cornell University. Born in Poland, the many honors awarded to him throughout his exceptional scientific career include the 1981 Nobel Prize in Chemistry, jointly with Kenichi Fukui, the National Medal of Science, the American Chemical Society's awards in three different subfields of chemistry, and numerous honorary degrees. The author of nearly six hundred articles for scientific journals, Hoffmann is a member of the National Academy of Sciences and of many foreign academies. His continual efforts to reach out to the general public include the production of a television course, *The World of Chemistry* (1989), and a monthly cabaret in New York City, *Entertaining Science*. As a writer, he carves out a space among science, poetry, and philosophy in *Chemistry Imagined* (1993), with artist Vivian Torrence, *The Same and Not the Same* (1979), and *Old Wine, New Flasks: Reflections on Science and Jewish Tradition* (1997), with Shira Leibowitz Schmidt; his essays have been collected in *Roald Hoffmann on the Philosophy, Art, and Science of Chem-*

istry (2012). He is also a poet and playwright, whose work has been published, produced, and translated widely: *The Metamict State* (1987), *Gaps and Verges: Poems* (1990), *Memory Effects: Poems* (1999), *Soliton: Poems* (2002), and the plays *Oxygen* (2001), with fellow chemist Carl Djerassi, and *Should've* (2011). The autobiographical play *Something That Belongs to You* had a German production in 2014.

EVELYN FOX KELLER is Professor Emerita of History and Philosophy of Science in the Program in Science, Technology and Society at MIT. After receiving her PhD in theoretical physics at Harvard University, she worked for several years at the interface of physics and biology before turning to the study of gender and science, and later, to the history and philosophy of science. Among her many books are the highly influential *A Feeling for the Organism: The Life and Work of Barbara McClintock* (1983), *Reflections on Gender and Science* (1985), *Refiguring Life: Metaphors of Twentieth-century Biology* (1995), *The Century of the Gene* (2000), and *Making Sense of Life: Explaining Biological Development with Models, Metaphors, and Machines* (2002). She is a member of the American Philosophical Society and the American Academy of Arts and Sciences, and the recipient of many awards, including a MacArthur Fellowship and a Guggenheim Fellowship, and numerous honorary degrees. Her most recent book is *The Mirage of a Space between Nature and Nurture* (2010).

JEAN-MICHEL RABATÉ is Professor of English at the University of Pennsylvania, where he has taught since 1992, and a managing editor of the *Journal of Modern Literature*. Born in France, Rabaté has authored or edited more than thirty books, among them *The Ghosts of Modernity* (1996), *James Joyce and the Politics of Egoism* (2001), *The Cambridge Companion to Jacques Lacan* (2003), *1913: The Cradle of Modernism* (2007), translated into Chinese in 2013. His most recent book is *Crimes of the Future* (2014). While his own books have been published in English and French, Rabaté is the translator from English into French of books by Daniel Gunn, Gabriel Josipovici, Ezra Pound, and Martin Amis. He is a Fellow of the American Academy of Arts and Sciences and a Senior Curator at the Slought Foundation in Philadelphia.

MARGERY ARENT SAFIR is Professor Emerita of Comparative Literature at The American University of Paris and was a member of the research group CRICCAL at the Sorbonne-Nouvelle. She is the author, co-author, and general editor of works on the leading figures of contemporary Latin American literature, including *Earth Tones: The Poetry of Pablo Neruda, with Manuel Durán*. She has also collaborated with distinguished scientists on the interactions of science and literature in such works as *Mélancolies du savoir* (1995) and the expanded *Melancholies of Knowledge: Literature in the Age of Science* (1999). Most recently, she was General Editor of *Robert Wilson from Within* (2011, English; 2012, French), a major reference work on the renowned *avant garde* theater artist. She is founder of The Arts Arena (2007), a nonprofit interdisciplinary initiative for the creative and performing arts, film, and issues of culture and society, and serves as its Executive/Artistic Director. The Arts Arena institutional partnerships include Columbia Global Centers | Europe, Curtis Institute of Music, Yale School of Art, and Yale School of Music.

www.ingramcontent.com/pod-product-compliance
Lightning Source LLC
LaVergne TN
LVHW101443200926
848483LV00001B/39
* 9 7 8 1 6 1 1 4 8 6 4 3 8 *